Mathematics Station Activities

for Common Core State Standards
Grade 7

WALCH EDUCATION®

1   2   3   4   5   6   7   8   9   10

ISBN 978-0-8251-7425-4

Copyright © 2011, 2013

J. Weston Walch, Publisher

Portland, ME 04103

www.walch.com

Printed in the United States of America

Table of Contents

Introduction

This revised edition of the *Mathematics Station Activities for Common Core State Standards, Grade 7* includes a collection of updated and improved station-based activities to provide students with opportunities to practice and apply the mathematical skills and concepts they are learning. It contains sets of activities that are tightly aligned to both the Mathematical Practices and the five Grade 7 Common Core Mathematics domains: Ratios and Proportional Relationships; The Number System; Expressions and Equations; Geometry; and Statistics and Probability. These enhancements have been carried out based on continuing refinement of Common Core implementation. You may use these activities in addition to direct instruction, or instead of direct instruction in areas where students understand the basic concepts but need practice. The Discussion Guide included with each set of activities provides an important opportunity to help students reflect on their experiences and synthesize their thinking. It also provides guidance for ongoing, informal assessment to inform instructional planning.

Implementation Guide

The following guidelines will help you prepare for and use the activity sets in this book.

Setting Up the Stations

Each activity set consists of four stations. Set up each station at a desk, or at several desks pushed together, with enough chairs for a small group of students. Place a card with the number of the station on the desk. Each station should also contain the materials specified in the teacher's notes, and a stack of student activity sheets (one copy per student). Place the required materials (as listed) at each station.

When a group of students arrives at a station, each student should take one of the activity sheets to record the group's work. Although students should work together to develop one set of answers for the entire group, each student should record the answers on his or her own activity sheet. This helps keep students engaged in the activity and gives each student a record of the activity for future reference.

Forming Groups of Students

All activity sets consist of four stations. You might divide the class into four groups by having students count off from 1 to 4. If you have a large class and want to have students working in small groups, you might set up two identical sets of stations, labeled A and B. In this way, the class can be divided into eight groups, with each group of students rotating through the "A" stations or "B" stations.

Assigning Roles to Students

Students often work most productively in groups when each student has an assigned role. You may want to assign roles to students when they are assigned to groups and change the roles occasionally. Some possible roles are as follows:

- Reader—reads the steps of the activity aloud
- Facilitator—makes sure that each student in the group has a chance to speak and pose questions; also makes sure that each student agrees on each answer before it is written down
- Materials Manager—handles the materials at the station and makes sure the materials are put back in place at the end of the activity
- Timekeeper—tracks the group's progress to ensure that the activity is completed in the allotted time
- Spokesperson—speaks for the group during the debriefing session after the activities

Timing the Activities

The activities in this book are designed to take approximately 15 minutes per station. Therefore, you might plan on having groups change stations every 15 minutes, with a two-minute interval for moving from one station to the next. It is helpful to give students a "5-minute warning" before it is time to change stations.

Since the activity sets consist of four stations, the above time frame means that it will take about an hour and 10 minutes for groups to work through all stations. If this is followed by a 20-minute class discussion as described on the next page, an entire activity set can be completed in about 90 minutes.

Guidelines for Students

Before starting the first activity set, you may want to review the following "ground rules" with students. You might also post the rules in the classroom.

- All students in a group should agree on each answer before it is written down. If there is a disagreement within the group, discuss it with one another.
- You can ask your teacher a question only if everyone in the group has the same question.
- If you finish early, work together to write problems of your own that are similar to the ones on the student activity sheet.
- Leave the station exactly as you found it. All materials should be in the same place and in the same condition as when you arrived.

Debriefing the Activities

After each group has rotated through every station, bring students together for a brief class discussion. At this time, you might have the groups' spokespersons pose any questions they had about the activities. Before responding, ask if students in other groups encountered the same difficulty or if they have a response to the question. The class discussion is also a good time to reinforce the essential ideas of the activities. The questions that are provided in the teacher's notes for each activity set can serve as a guide to initiating this type of discussion.

You may want to collect the student activity sheets before beginning the class discussion. However, it can be beneficial to collect the sheets afterward so that students can refer to them during the discussion. This also gives students a chance to revisit and refine their work based on the debriefing session.

Standards Correlations

The standards correlations below and on the next page support the implementation of the Common Core State Standards. This book includes station activity sets for the Common Core domains of Ratios and Proportional Relationships; The Number System; Expressions and Equations; Geometry; and Statistics and Probability. The table that follows provides a listing of the available station activities organized by Common Core standard.

The left column lists the standard codes. The first number of the code represents the grade level. The grade number is followed by the initials of the Common Core domain name, which is then followed by the standard number. The middle column of the table lists the title of the station activity set that corresponds to the standard, and the right column lists the page number where the station activity set can be found.

Standard	Set title	Page number
7.RP.1.	Ratios and Proportions	1
7.RP.2.	Ratios and Proportions	1
7.RP.2.	Analyzing and Describing Relationships	21
7.RP.2.	Graphing Relationships	28
7.RP.2.	Representing Patterns and Relationships	37
7.RP.3.	Ratios and Proportions	1
7.RP.3.	Problem Solving with Rational Numbers	14
7.EE.3.	Problem Solving with Rational Numbers	14
7.NS.1.	Adding and Subtracting Rational Numbers	45
7.NS.2.	Multiplying and Dividing Rational Numbers	53
7.EE.3.	Multi-Step Real-Life Problems with Rational Numbers	61
7.EE.4.	Using Variables to Construct Equations and Inequalities	69
7.EE.4.	Solving Inequalities	76

(continued)

Standards Correlations

Standard	Set title	Page number
7.G.1.	Similarity and Scale	85
7.G.2.	Similarity and Scale	85
7.G.3.	Sketching, Modeling, and Describing 3-D Figures	93
7.SP.5.	Collecting, Organizing, and Analyzing Data	100
7.SP.5.	Theoretical Probability	108
7.SP.6.	Collecting, Organizing, and Analyzing Data	100
7.SP.6.	Theoretical Probability	108
7.SP.6.	Experimental Probability	115
7.SP.7.	Collecting, Organizing, and Analyzing Data	100
7.SP.7.	Experimental Probability	115
7.SP.8.	Collecting, Organizing, and Analyzing Data	100

Materials List

Station Sets

- 2 bags (to hold cubes/blocks/etc.)
- 2 spinners (one divided evenly into red, green, blue, and yellow sections; the other divided evenly into black and white sections)
- 3 blocks (red, green, and blue)
- 3 marshmallows
- box of toothpicks
- clay
- colored cubes (one each of red, blue, and green; two yellow)
- graph paper
- integer chips
- list of 100 randomly generated numbers (1–9)
- marbles (24 green, 16 yellow)
- model car with scale factor
- *optional:* tape measure; water; separate containers to hold 1 cup, 1 quart, and 1 gallon of water
- plastic knives
- rectangular prism
- tiles or small pieces of paper (several small square tiles; several small round tiles; 8 large blue algebra tiles; 20 small yellow algebra tiles)

Class Sets

- calculators
- protractors
- rulers

Ongoing Use

- index cards (prepared according to specifications in teacher notes for many of the station activities)
- number cubes (several numbered 1–6; one numbered 4, 2, 2, 1, 3, 3; one numbered 6, 1, 3, 5, 4, 8)
- pencils
- pennies
- scrap/scratch paper

Ratios and Proportional Relationships

Set 1: Ratios and Proportions

Goal: To provide opportunities for students to develop concepts and skills related to unit conversion, finding percents, simplifying algebraic ratios, and solving algebraic proportions

Common Core State Standards

Ratios and Proportional Relationships

Analyze proportional relationships and use them to solve real-world and mathematical problems.

7.RP.1. Compute unit rates associated with ratios of fractions, including ratios of lengths, areas and other quantities measured in like or different units.

7.RP.2. Recognize and represent proportional relationships between quantities.

  a. Decide whether two quantities are in a proportional relationship, e.g., by testing for equivalent ratios in a table or graphing on a coordinate plane and observing whether the graph is a straight line through the origin.

  b. Identify the constant of proportionality (unit rate) in tables, graphs, equations, diagrams, and verbal descriptions of proportional relationships.

  c. Represent proportional relationships by equations.

7.RP.3. Use proportional relationships to solve multi-step ratio and percent problems. Examples: simple interest, tax, markups and markdowns, gratuities and commissions, fees, percent increase and decrease, percent error.

Student Activities Overview and Answer Key

Station 1

Students will be given 12 index cards with pairs of equivalent units of measurement written on them. They will work together to match the cards that are an equivalent unit of measurement. Then they will perform unit conversion.

Answers

1. 10 mm = 1 cm; 12 in. = 1 ft; 3 ft = 1 yd; 2 pints = 1 quart; 4 quarts = 1 gallon; 1 ton = 2,000 pounds

2. 8 pints in a gallon; 2 pints = 1 quart and 4 quarts = 1 gallon, so 2(4) = 8 pints

3. 18 inches; 1/2 yard = 1.5 feet and 12 inches = 1 foot, so 12(1.5) = 18 inches

4. 5,000 pounds

5. 850 mm

6. 13.5 feet

7. 3 quarts = 0.75 gallons

8. Answers will vary. Possible answers include: cooking, when modifying recipes for more or fewer people; carpentry, when creating custom-size cabinetry

Station 2

Students will be given a calculator to help them solve the problems. They work as a group to solve real-world applications of unit conversions.

Answers

1. His friend measures temperature in Celsius, and Evan measures it in Fahrenheit. $F = 95°$

2. $P = 36.67$ yards; $P = 1,320$ inches, $A = 77.78$ yds^2; $A = 100,800$ in^2

3.

	Feet	Yards	Meters	Time
Tim	300	100	91.44	12 seconds
Jeremy	400	133.33	121.95	12 seconds
Martin	229.66	76.55	70	12 seconds

Jeremy, Tim, Martin; Tim = 25 feet/sec, Jeremy = 33.33 feet/sec; Martin = 19.14 feet/sec

Station 3

Students will be given a bag containing 24 green marbles and 16 yellow marbles. They will use the marbles to create ratios and percents. They will then solve percent problems.

Answers

1. Answers will vary. Possible answers include: green = 1; yellow = 7; total = 8. Find 1/8 = 0.125 = 12.5%; 12.5% were green. Subtract 12.5% from 100% to get 87.5% or 7/8 = 87.5%; 87.5% were yellow.

2. There are 40 marbles so 24/40 = 60% green marbles; 100% − 60% = 40% or 16/40 = 40%

3. 9 yellow marbles; student drawings should depict 9 yellow marbles and 12 green marbles.

4. 24(1/4) = 6 or 24(0.25) = 6

5. 17(2/1) = 34 or 17(2.0) = 34

6. 10(14) = 140 in^2; increased dimensions by 200% then found the area of the photograph

Station 4

Students will be given 8 large blue algebra tiles and 20 small yellow algebra tiles. Students visually depict ratios and proportions with the algebra tiles. They then solve proportions for a specified variable including a real-world application.

Answers

1. $\dfrac{8 \text{ blue}}{20 \text{ yellow}} = \dfrac{2}{5}$

2. $\dfrac{2 \text{ blue}}{3 \text{ yellow}} = \dfrac{4 \text{ blue}}{6 \text{ yellow}}$

3. $8/20 = x/100$, so $x = 40$ blue

4. $8/20 = x/15$, so $x = 6$ blue

5. $x = 4$

6. $x = 40$

7. $\dfrac{\text{blue}}{\text{yellow}} = \dfrac{6}{10} = \dfrac{3}{5}$

   Let x = number of blue pencils and $24 - x$ = number of yellow pencils.

   $\dfrac{3}{5} = \dfrac{x}{(24 - x)}$, so $x = 9$ blue pencils and $24 - x = 15$ yellow pencils

Materials List/Setup

Station 1	12 index cards with the following written on them:
	10 millimeters, 12 inches, 3 feet, 2 pints, 4 quarts, 1 ton, 1 centimeter, 1 foot, 1 yard, 1 quart, 1 gallon, 2,000 pounds
Station 2	calculator
Station 3	24 green marbles; 16 yellow marbles
Station 4	8 large blue algebra tiles; 20 small yellow algebra tiles

Discussion Guide

To support students in reflecting on the activities and to gather some formative information about student learning, use the following prompts to facilitate a class discussion to "debrief" the station activities.

Prompts/Questions

1. How do you perform unit conversion?

2. When would you use unit conversion in the real world?

3. What are two ways to find the percent of a number?

4. What is a ratio?

5. How do you know if two ratios are equivalent?

6. What is a proportion?

7. When would you use ratios and proportions in the real world?

Think, Pair, Share

Have students jot down their own responses to questions, then discuss with a partner (who was not in their station group), and then discuss as a whole class.

Suggested Appropriate Responses

1. Use ratios and proportions to convert units.

2. Answers will vary. Possible answers include: creating scale models of buildings; using the metric system instead of U.S. customary units; converting Celsius to degrees Fahrenheit and vice versa

3. Multiply the number by a decimal or fraction that represents the percentage.

4. A ratio is a comparison of two numbers by division.

5. Two ratios are equivalent if, when simplified, they are equal.

6. A proportion is when two ratios are set equal to each other.

7. Answers will vary. Possible answers include: enlarging photos; scale models; modifying quantities of ingredients in a recipe

Possible Misunderstandings/Mistakes

- Not keeping track of units and using incorrect unit conversions
- Not recognizing that terms must have the same units in order to compare them
- Setting up proportions with one of the ratios written with the incorrect numbers in the numerator and denominator
- Not recognizing simplified forms of ratios in order to find equivalent ratios

Ratios and Proportional Relationships
Set 1: Ratios and Proportions

Station 1

You will be given 12 index cards with the following written on them:

> 10 millimeters, 12 inches, 3 feet, 2 pints, 4 quarts, 1 ton, 1 centimeter, 1 foot, 1 yard, 1 quart, 1 gallon, 2,000 pounds

Shuffle the index cards and deal a card to each student in your group until all the cards are gone. As a group, show your cards to each other and match the cards that are an equivalent unit of measurement.

1. Write your answers on the lines below. The first match is shown:

   10 mm = 1 cm

   _____   _____

   _____   _____

2. Find the number of pints in a gallon. Explain how you can use your answers in problem 1 to find the number of pints in a gallon.

3. Find the number of inches in half of a yard. Explain how you can use your answers in problem 1 to find the number of inches in half of a yard.

continued

Ratios and Proportional Relationships

Set 1: Ratios and Proportions

Perform the following unit conversions by filling in the blanks.

4. 2.5 tons = _____ pounds

5. 85 cm = _____ mm

6. 4.5 yd = _____ ft

7. 6 pints = _____ quarts = _____ gallons

8. When would you use unit conversions in the real world?

Ratios and Proportional Relationships

Set 1: Ratios and Proportions

Station 2

You will be given a calculator to help you solve the problems. Work as a group to solve these real-world applications of unit conversions.

1. Evan has a friend in England. His friend said the temperature was very hot at 35°. Evan thought he heard his friend incorrectly since 35° is cold. What caused his misunderstanding?

   (*Hint*: $C = (F - 32)\dfrac{5}{9}$)

   Find the equivalent temperature in the United States that would make the claim of Evan's friend valid. Write your answer in the space below.

2. Anna is going to build a patio. She wants the patio to be 20 feet by 35 feet. What is the perimeter of the patio in yards?

   What is the perimeter of the patio in inches?

   What is the area of the patio in yards?

continued

Ratios and Proportional Relationships
Set 1: Ratios and Proportions

What is the area of the patio in inches?

3. Tim claims he can run the 100-yard dash in 12 seconds. Jeremy claims he can run 400 feet in 12 seconds. Martin claims he can run 70 meters in 12 seconds. (*Hint*: 1 yard = 0.9144 meters and 1 yard = 3 feet.)

Fill in the table below to create equivalent units of measure.

	Feet	Yards	Meters	Time (seconds)
Tim				
Jeremy				
Martin				

List the three boys in order of fastest to slowest:

How fast did each boy run in feet/second?

Ratios and Proportional Relationships
Set 1: Ratios and Proportions

Station 3

You will be given a bag containing 24 green marbles and 16 yellow marbles. You will use the marbles to create ratios and percents. You will then solve percent problems. Work together as a group to solve the following problems.

1. Shake the bag of green and yellow marbles so that the colors are mixed. Have each group member select 2 marbles from the bag without looking. Group all your marbles together by color.

   How many green marbles did you draw? _____

   How many yellow marbles did you draw? _____

   What was the total number of marbles drawn? _____

   How can you determine the percentage of marbles that were green?

   Find the percentage of marbles you drew that were green.

   Name two ways you can find the percentage of marbles you drew that were yellow.

   Find the percentage of marbles you drew that were yellow.

2. Take all the marbles out of the bag. How can you determine what percentage of all the marbles are green?

   How can you determine what percentage of all the marbles are yellow?

continued

Ratios and Proportional Relationships
Set 1: Ratios and Proportions

3. Place 12 green marbles on the table. How many yellow marbles do you need to have 75% as many yellow marbles on the table?

   Draw a picture of the number of green marbles and yellow marbles you have placed on the table.

4. Use equations to show two ways you can find 25% of 24.

5. Use equations to show two ways you can find 200% of 17.

6. Real-world application: Bryan is a photographer. He has a 5 inch by 7 inch photo that he wants to enlarge by 200%. What is the area of the new photo? Explain your answer in the space below.

Ratios and Proportional Relationships
Set 1: Ratios and Proportions

Station 4

You will be given 8 large blue algebra tiles and 20 small yellow algebra tiles. Work as a group to arrange the algebra tiles so they visually depict the ratio of blue to yellow algebra tiles.

1. What is this ratio? _____

Rearrange the tiles to visually depict the following ratios:

$$\frac{2 \text{ blue}}{3 \text{ yellow}} \qquad \frac{1 \text{ blue}}{10 \text{ yellow}} \qquad \frac{4 \text{ blue}}{6 \text{ yellow}} \qquad \frac{1 \text{ blue}}{1 \text{ yellow}}$$

2. Which ratios are equivalent ratios? Explain your answer.

3. Keeping the same ratio of yellow to blue tiles, if there were 100 yellow algebra tiles, how many blue algebra tiles would there be? Use a proportion to solve this problem. Show your work in the space below. (*Hint*: A proportion is two ratios that are equal to each other.)

4. Keeping the same ratio of yellow to blue tiles, if there were 15 yellow algebra tiles, how many blue algebra tiles would there be? Use a proportion to solve this problem. Show your work in the space below.

continued

Ratios and Proportional Relationships

Set 1: Ratios and Proportions

Work together to solve the following proportions for the variable.

5. $\dfrac{2}{7} = \dfrac{x}{14}$; $x =$

6. $\dfrac{8}{x} = \dfrac{2}{10}$; $x =$

Use the following information to answer problem 7:

Allison has 6 blue pencils and 10 yellow pencils. Sadie has 24 pencils that are either blue or yellow. The ratio of blue pencils to yellow pencils is the same for both Allison and Sadie.

7. How many blue pencils and yellow pencils does Sadie have? Show your work in the space below by setting up a proportion using a variable, x.

Ratios and Proportional Relationships

Goal: To provide opportunities for students to solve problems involving rational numbers

Common Core State Standards

Ratios and Proportional Relationships

Analyze proportional relationships and use them to solve real-world and mathematical problems.

7.RP.3. Use proportional relationships to solve multistep ratio and percent problems. Examples: simple interest, tax, markups and markdowns, gratuities and commissions, fees, percent increase and decrease, percent error.

Expressions and Equations

Solve real-life and mathematical problems using numerical and algebraic expressions and equations.

7.EE.3. Solve multi-step real-life and mathematical problems posed with positive and negative rational numbers in any form (whole numbers, fractions, and decimals), using tools strategically. Apply properties of operations to calculate with numbers in any form; convert between forms as appropriate; and assess the reasonableness of answers using mental computation and estimation strategies.

Student Activities Overview and Answer Key

Station 1

Students work together to solve a real-world problem involving percent discounts. Students are encouraged to brainstorm appropriate problem-solving strategies and to explain their solution process once all students in the group agree upon the solution.

Answers: The regular price was $60.

Possible strategies: Work backward.

Possible steps: Work backward to find that the price before Wednesday's discount was $36, before Tuesday's discount was $45, and before Monday's discount was $60.

Station 2

Students work together to solve a real-world problem involving coins. This requires students to do arithmetic with rational numbers in the form of decimals. After reading the problem, students brainstorm possible problem-solving strategies. Students are encouraged to make sure everyone in the group agrees on the solution. Then students explain their solution process.

Answers: 15 quarters and 6 nickels

Possible strategies: Guess and check various possibilities; make an organized list of all possible combinations of quarters and nickels.

Possible steps: Try various combinations of quarters and nickels that add up to a total of 21 coins. Find the total value of the coins. Adjust the guesses to find a combination for which the total value of the coins is $4.05.

Station 3

Students work together to solve a real-world problem involving fractions. Students are encouraged to brainstorm different strategies that may be used to solve the problem and to make sure everyone in the group agrees on the solution. (The problem is especially well-suited to drawing a diagram.) Students write an explanation of their solution process.

Answers: 24 posts

Possible strategies: Draw a diagram.

Possible steps: Draw a rectangle and mark the dimensions of the plot. Mark the posts around the edge of the plot, noting that every four posts will span 9 feet of the perimeter. Count the total number of posts needed.

Station 4

Students work together to solve a real-world problem involving the ratio of girls to boys at a school. Students can solve the problem in many ways, including using proportional reasoning. After reading the problem, students brainstorm possible problem-solving strategies. Students are encouraged to make sure everyone in the group agrees on the solution. Then students explain their solution process.

Answers: 51 girls and 34 boys

Possible strategies: Guess and check various possibilities; find different combinations of girls and boys in which the number of girls is $1\frac{1}{2}$ times the number of boys, and increase or decrease the numbers of girls and boys proportionally until the total number of students is 85.

Possible steps: Try various combinations of girls and boys, such as 30 girls and 20 boys, in which the number of girls is $1\frac{1}{2}$ times the number of boys. Continue until the total number of students is exactly 85. Alternatively, find combinations of numbers that have a sum of 85 and adjust the numbers until they differ by a factor of $1\frac{1}{2}$.

Materials List/Setup

No materials needed

Discussion Guide

To support students in reflecting on the activities and to gather some formative information about student learning, use the following prompts to facilitate a class discussion to "debrief" the station activities.

Prompts/Questions

1. What are some different problem-solving strategies you can use to help you solve real-world problems?

2. How do you perform arithmetic operations involving percents when you solve a problem?

3. How do you decide when to use a diagram to help you solve a problem?

4. How can you check your answer to a real-world problem?

Think, Pair, Share

Have students jot down their own responses to questions, then discuss with a partner (who was not in their station group), and then discuss as a whole class.

Suggested Appropriate Responses

1. Make a table, guess and check, look for a pattern, work backward, draw a diagram, use physical objects to model the problem, etc.

2. First convert the percents to decimals.

3. Use a diagram when the problem involves geometric shapes or relationships, or when the information can be represented on a number line.

4. Reread the problem using the answer in place of the unknown quantity or quantities. Check to see if the numbers work out correctly throughout the problem.

Possible Misunderstandings/Mistakes

- Incorrectly converting percents to decimals

- Using the amount of a discount as the sale price (e.g., writing $25 as the sale price of a $100 item that is discounted by 25%)

- Incorrectly using decimals to write the value of coins (e.g., writing 5 or 0.5 for the value of a nickel instead of 0.05)

Ratios and Proportional Relationships
Set 2: Problem Solving with Rational Numbers

Station 1

At this station, you will work with other students to solve this real-life problem.

> A department store is having a special sale. On Monday, the regular price of a dress is marked down by 25%. On Tuesday, the dress is marked down by 20% of its current price. On Wednesday, the dress is marked down by 25% of its current price. After all the discounts, the final price of the dress is $27. What was the regular price of the dress?

Work with other students to discuss the problem. Brainstorm strategies you might use to solve the problem. Write the strategies below.

_____

_____

_____

_____

_____

Solve the problem. When everyone agrees on the answer, write it below.

_____

Explain the steps you used to solve the problem.

_____

_____

_____

_____

_____

Ratios and Proportional Relationships
Set 2: Problem Solving with Rational Numbers

Station 2

At this station, you will work with other students to solve this real-life problem.

Latrell has a jar containing quarters and nickels. There are 21 coins in the jar altogether. The total value of the coins is $4.05. How many quarters and how many nickels are in the jar?

Work with other students to discuss the problem. Brainstorm strategies you might use to solve the problem. Write the strategies below.

_____

_____

_____

_____

_____

Solve the problem. When everyone agrees on the answer, write it below.

_____

Explain the steps you used to solve the problem.

_____

_____

_____

_____

_____

_____

Ratios and Proportional Relationships
Set 2: Problem Solving with Rational Numbers

Station 3

At this station, you will work with other students to solve this real-life problem.

Anna wants to put a fence around a rectangular garden plot that measures 9 feet by 18 feet. She wants to place posts at the corners of the plot and every $2\frac{1}{4}$ feet around the perimeter of the plot. How many posts does she need?

Work with other students to discuss the problem. Brainstorm strategies you might use to solve the problem. Write the strategies below.

_____

_____

_____

_____

_____

Solve the problem. When everyone agrees on the answer, write it below.

_____

Explain the steps you used to solve the problem.

_____

_____

_____

_____

_____

_____

Ratios and Proportional Relationships
Set 2: Problem Solving with Rational Numbers

Station 4

At this station, you will work with other students to solve this real-life problem.

> At Harriet Tubman Middle School, there are 85 students in the 7th grade. The number of girls is $1\frac{1}{2}$ times the number of boys. How many girls and how many boys are in the 7th grade?

Work with other students to discuss the problem. Brainstorm strategies you might use to solve the problem. Write the strategies below.

_____

_____

_____

_____

_____

Solve the problem. When everyone agrees on the answer, write it below.

_____

Explain the steps you used to solve the problem.

_____

_____

_____

_____

_____

_____

Ratios and Proportional Relationships

Set 3: Analyzing and Describing Relationships

Goal: To provide opportunities for students to develop concepts and skills related to analyzing and describing relationships

Common Core State Standards

Ratios and Proportional Relationships

Analyze proportional relationships and use them to solve real-world and mathematical problems.

7.RP.2. Recognize and represent proportional relationships between quantities.

b. Identify the constant of proportionality (unit rate) in tables, graphs, equations, diagrams, and verbal descriptions of proportional relationships.

c. Represent proportional relationships by equations.

Student Activities Overview and Answer Key

Station 1

In this activity, students work together to translate verbal phrases into algebraic expressions. Students are given information about a library's late fees. They translate this information into algebraic expressions that can be used to calculate the late fees. Students explain their work and use their expressions to calculate late fees.

Answers

1. $0.75, $1, $1.75

2. $0.5 + 0.25d$

3. The fee is $0.25 per day (which can be written as $0.25d$) plus $0.50, so the total fee is $0.5 + 0.25d$.

4. $2 + 0.5d$

5. $1.5d$

6. Evaluate $1.5d$ for $d = 8$, so the fee is $1.5(8) = 12.

Station 2

Students work together to analyze a pattern based on the cost of renting DVDs. They work together to complete and extend a table of values. Then students write an expression that can be used to find the cost of renting any number of DVDs. Finally, they use the expression to find the cost of renting 15 DVDs.

Answers

The missing table values are $17, $20, and $23.

1. $26, $29

2. Possible explanation: Extend the table by adding $3 for each additional DVD.

3. $8 + 3m$

4. Possible explanation: The cost of renting m DVDs is $8 plus $3 per DVD, which is $8 + 3m$.

5. The cost of renting 15 DVDs is $8 + 3(15) = \$53$.

Station 3

Students work together to analyze and extend a pattern that is provided in the form of a table. The table shows the (decreasing) number of contestants on a TV show from one week to the next. Students are asked to write and use an expression that describes the number of contestants during any given week.

Answers

1. 30 contestants

2. Extend the table to the 8th week by subtracting 6 contestants each week.

3. $72 - 6(x - 1)$ or $78 - 6x$

4. In week x, a total of 6 times $x - 1$ contestants have been eliminated, so subtract $6(x - 1)$ from 72.

5. In week 10, the number of contestants is $72 - 6(9) = 72 - 54 = 18$.

Station 4

At this station, students analyze and describe a relationship that is given in a graph. The graph shows the cost of renting ice skates for various numbers of hours. Students work together to predict the cost for a number of hours that is not shown on the graph. They also write and use an expression that gives the cost for h hours.

Answers

1. $3, $5, $7

2. $13

3. The cost is 2 times the number of hours plus $1. (Students may also extend the graph.)

4. $2h + 1$

5. To find the cost for 8 hours, let $h = 8$ in the expression to get $2(8) + 1 = \$17$.

Materials List/Setup

No materials needed

Discussion Guide

To support students in reflecting on the activities and to gather some formative information about student learning, use the following prompts to facilitate a class discussion to "debrief" the station activities.

Prompts/Questions

1. What are some strategies you can use to help translate verbal phrases into algebraic expressions?

2. What are some strategies you can use to help extend a table of data?

3. Why is it useful to translate information from tables or graphs into algebraic expressions?

4. How can you use an expression that gives the cost of n apples to find the cost of 7 apples?

Think, Pair, Share

Have students jot down their own responses to questions, then discuss with a partner (who was not in their station group), and then discuss as a whole class.

Suggested Appropriate Responses

1. Use verbal clues about operations. For example, the word plus means "addition," which will be represented by a "+" symbol in the algebraic expression.

2. Look for a pattern in the sequence of numbers. For example, a fixed number may be added or subtracted each time you go from one number to the next.

3. The algebraic expression is more general. You can use it to calculate additional values that are not shown in the table or graph.

4. Evaluate the expression for $n = 7$.

Possible Misunderstandings/Mistakes

- Assuming that all patterns may be described multiplicatively (e.g., in the form $y = kx$)

- Incorrectly representing operations when writing an expression (e.g., writing $n + 2$ instead of $2n$)

- Multiplying the variable by an incorrect constant (e.g., writing $8m + 3$ instead of $8 + 3m$)

Ratios and Proportional Relationships
Set 3: Analyzing and Describing Relationships

Station 1

At this station, you will work with other students to translate words into algebra. Use the following information.

Midtown Library	
Late Fees	
Books	$0.50, plus $0.25 for each day that the book is late
CDs	$2, plus $0.50 for each day that the CD is late
DVDs	$1.50 per day that the DVD is late

Work together to complete the following. When everyone agrees on an answer, write it in the space provided.

1. What is the fee for a book that is 1 day late? 2 days late? 5 days late? _____

   _____

2. Write an expression that gives the fee for a book that is d days late. _____

3. Explain how you came up with your answer to problem 2. _____

   _____

   _____

   _____

4. Write an expression that gives the fee for a CD that is d days late. _____

5. Write an expression that gives the fee for a DVD that is d days late. _____

6. Show how to use an expression you wrote to find the fee for a DVD that is 8 days late.

Ratios and Proportional Relationships
Set 3: Analyzing and Describing Relationships

Station 2

At this station, you will explore a pattern based on prices.

Renting DVDs costs $8 for a membership, plus $3 for each DVD rented. The table shows the cost of renting different numbers of DVDs.

Work with other students to complete the table.

Number of DVDs	1	2	3	4	5
Cost	$11	$14			

Work together to complete the following. When everyone agrees on an answer, write it in the space provided.

1. Predict the cost of renting 6 DVDs and 7 DVDs. _____

2. Explain how you made these predictions. _____

   _____

   _____

   _____

3. Write an expression that gives the cost of renting m DVDs. _____

4. Explain how you came up with the expression. _____

   _____

   _____

   _____

5. Show how you can use your expression to find the cost of renting 15 DVDs.

Ratios and Proportional Relationships
Set 3: Analyzing and Describing Relationships

Station 3

At this station, you will explore a pattern based on a table of data.

A television show begins with 72 contestants during the first week. Each week, 6 contestants are eliminated from the show. The table below shows the number of contestants during the first few weeks of the show.

Week	1	2	3	4
Number of contestants	72	66	60	54

Work together to complete the following. When everyone agrees on an answer, write it in the space provided.

1. Predict the number of contestants on the show during the 8th week. _____

2. Explain how you made this prediction. _____

    _____

    _____

    _____

3. Write an expression that gives the number of contestants during week x of the show.

    _____

4. Explain how you came up with the expression. _____

    _____

    _____

    _____

5. Show how you can use the expression to find the number of contestants during the 10th week of the show.

NAME:

Ratios and Proportional Relationships
Set 3: Analyzing and Describing Relationships

Station 4

At this station, you will work with other students to analyze data in a graph.

The graph to the right shows the cost of renting ice skates for various numbers of hours.

Work together to solve the following problems. When everyone agrees on an answer, write it in the space provided.

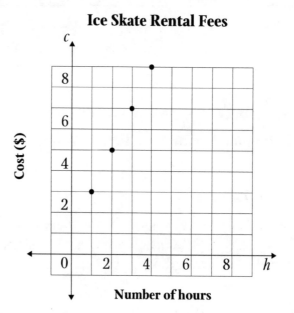

Ice Skate Rental Fees

1. What does it cost to rent ice skates for 1 hour? 2 hours? 3 hours?

   _____

2. Predict the cost of renting skates for 6 hours. _____

3. Explain how you made this prediction. _____

   _____

   _____

   _____

4. Write an expression that gives the cost of renting skates for *h* hours. _____

5. Show how you can use your expression to find the cost of renting skates for 8 hours.

Ratios and Proportional Relationships

Set 4: Graphing Relationships

Goal: To provide opportunities for students to develop concepts and skills related to coordinate graphing

Common Core State Standards

Ratios and Proportional Relationships

Analyze proportional relationships and use them to solve real-world and mathematical problems.

7.RP.2.   Recognize and represent proportional relationships between quantities.

    a.   Decide whether two quantities are in a proportional relationship, e.g., by testing for equivalent ratios in a table or graphing on a coordinate plane and observing whether the graph is a straight line through the origin.

    b.   Identify the constant of proportionality (unit rate) in tables, graphs, equations, diagrams, and verbal descriptions of proportional relationships.

    c.   Represent proportional relationships by equations.

Student Activities Overview and Answer Key

Station 1

Students work together to complete a table for the function $y = x + 4$. They use the completed table to write ordered pairs that satisfy the function, and then they work together to plot these points on a coordinate plane. Once students have plotted the points, they draw and describe the complete graph.

Answers

1.

x	y
−5	−1
−1	3
0	4
2	6
3	7

2.   (−5, −1), (−1, 3), (0, 4), (2, 6), (3, 7)

3–5.

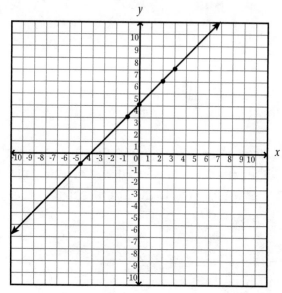

6. The graph is an upward-sloping straight line.

Station 2

In this activity, students are given a table of values. They work together to plot these points and look for patterns. (The points all satisfy the relationship $y = -2x + 3$, so they all lie on a straight line.) Using the graph of plotted points, students name an additional point that they think satisfies the same relationship, and they justify their choice.

Answers

1. (–5, 13), (–2, 7), (0, 3), (3, –3), (4, –5)

2–3.

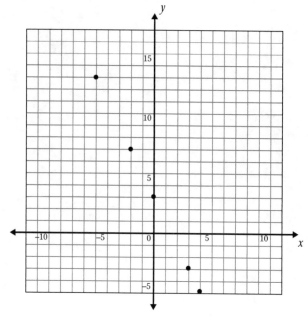

4. The points all lie on a straight line.

5. Possible answer: (–3, 9) or (2, –1).

6. The point lies along the line determined by the other points.

Station 3

Students work together to explore a pattern made from tiles. They complete a table showing the relationship between the stage of the pattern and the number of tiles needed. Then they plot the points (which lie on the line $y = 3x + 1$) and use the graph to predict the number of tiles that would be needed to make Stage 8 of the pattern.

Answers

1.

Stage (x)	1	2	3	4	5
Number of tiles (y)	4	7	10	13	16

2. (1, 4), (2, 7), (3, 10), (4, 13), (5, 16)

3–4.
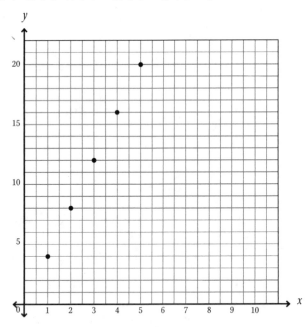

5. 25 tiles

6. Possible explanation: The plotted points all lie on a line. Continue the pattern of points until the x-value is 8. The point is (8, 25).

Station 4

Students work together to explore and graph a relationship based on a real-life situation. First, students complete a table of values that show the temperature of a freezer at various times. Then they work together to plot the points. They use their graph to predict the number of hours it will take for the temperature of the freezer to be 2°F.

Answers

1.

Hour (x)	0	1	4	7	9
Temperature, °F (y)	20	17	8	–1	–7

2. (0, 20), (1, 17), (4, 8), (7, –1), (9, –7)

3–4.

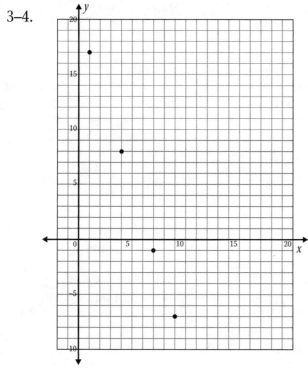

5. 6 hours

6. The plotted points all line on a straight line. Find the point along this line where the y-value is 2 (this corresponds to 2°F). The point is (6, 2).

Materials List/Setup

Station 1   graph paper; ruler

Station 2   graph paper; ruler

Station 3   small square tiles; graph paper; ruler

Station 4   graph paper; ruler

Discussion Guide

To support students in reflecting on the activities and to gather some formative information about student learning, use the following prompts to facilitate a class discussion to "debrief" the station activities.

Prompts/Questions

1. How do you set up a coordinate plane on a sheet of graph paper?

2. How do you plot an ordered pair on a coordinate plane?

3. How do you get ordered pairs from a table of values?

4. Is the ordered pair (2, 5) the same as the ordered pair (5, 2)? Why or why not?

Think, Pair, Share

Have students jot down their own responses to questions, then discuss with a partner (who was not in their station group), and then discuss as a whole class.

Suggested Appropriate Responses

1. Use a ruler to draw a straight line (x-axis). Draw another line at a right angle to the first line (y-axis). Label the intersection as the origin, (0, 0).

2. First plot the x-value by moving along the x-axis by the given number of units (right if the x-value is positive, left if it is negative). Then move up or down from this point to plot the y-value (up if the y-value is positive, down if it is negative).

3. In each row or column, take the x-value and the corresponding y-value, in that order, to form one ordered pair.

4. No. In (2, 5), the x-value is 2 and the y-value is 5. In (5, 2), the x-value is 5 and the y-value is 2. These are different points on the coordinate plane.

Possible Misunderstandings/Mistakes

*   Plotting points incorrectly due to reversing the x- and y-values (e.g., plotting (3, 4) rather than (4, 3))

*   Plotting points incorrectly due to mislabeling or incorrectly calibrating the scale on the x- or y-axis

*   Incorrectly reading the ordered pairs from a table of values

Ratios and Proportional Relationships
Set 4: Graphing Relationships

Station 1

At this station, you will work together to make a table and graph for the equation $y = x + 4$.

The equation $y = x + 4$ states that the value of y is equal to the value of x plus 4.

1. Work with other students to complete the table of values.

x	y
−5	
−1	
0	
2	
3	

2. Write the ordered pairs of values, (x, y), from your table.

   _____

3. Set up a coordinate plane on a sheet of graph paper. Use a ruler to draw the x-axis and the y-axis.

4. Plot the ordered pairs from your table. Work together to make sure that all the points are plotted correctly.

5. Use the points to draw the complete graph of the equation $y = x + 4$.

6. Describe the graph of the equation $y = x + 4$. _____

   _____

   _____

   _____

   _____

   _____

Ratios and Proportional Relationships
Set 4: Graphing Relationships

Station 2

At this station, you will plot points from a table and use your graph to make predictions.

The table shows a set of values. Work with other students to explore how the x- and y-values are related.

x	y
−5	13
−1	7
0	3
3	−3
4	−5

1. Write the ordered pairs of values, (x, y), from the table.

   _____

2. Set up a coordinate plane on a sheet of graph paper. Use a ruler to draw the x-axis and the y-axis.

3. Plot the ordered pairs. Work together to make sure that all the points are plotted correctly.

4. Describe what you notice about the points. _____

   _____

   _____

   _____

5. Add a new point to the graph that you think has the same relationship between the x-value and y-value. What are the coordinates of the point?

   _____

6. Explain how you decided on this point. _____

   _____

   _____

   _____

Ratios and Proportional Relationships
Set 4: Graphing Relationships

Station 3

At this station, you will use a table and graph to explore a pattern. You can also use tiles to build the pattern.

Andre is using tiles to make a pattern for a walkway in his garden. Here are the first three stages in his pattern.

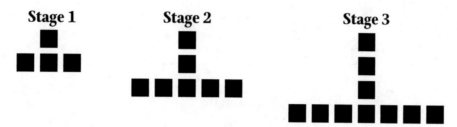

1. Complete the table.

Stage (x)	1	2	3	4	5
Number of tiles (y)					

2. Write the ordered pairs of values, (x, y), from the table.

   _____

3. Set up a coordinate plane on a sheet of graph paper. Use a ruler to draw the x-axis and the y-axis.

4. Plot the ordered pairs. Work together to make sure that all the points are plotted correctly.

5. Use your graph to predict the number of tiles that are needed at Stage 8.

   _____

6. Explain how you made this prediction. _____

   _____

   _____

   _____

Ratios and Proportional Relationships
Set 4: Graphing Relationships

Station 4

Work together to analyze a table of data from a real-life situation.

A scientist is doing an experiment with bacteria in a freezer. She sets the temperature of the freezer to 20°F. Each hour, she decreases the temperature by 3°F.

1. Complete the table.

Hour (x)	0	1	4	7	9
Temperature, °F (y)					

2. Write the ordered pairs of values, (x, y), from the table.

    _____

3. Set up a coordinate plane on a sheet of graph paper. Use a ruler to draw the x-axis and the y-axis.

4. Plot the ordered pairs. Work together to make sure that all the points are plotted correctly.

5. Use your graph to predict the number of hours it will take for the temperature in the freezer to reach 2°F.
    _____ hours

6. Explain how you made this prediction. _____

    _____

    _____

    _____

Ratios and Proportional Relationships

Set 5: Representing Patterns and Relationships

Goal: To provide opportunities for students to develop concepts and skills related to representing patterns and relationships

Common Core State Standards

Ratios and Proportional Relationships

7.RP.2. Recognize and represent proportional relationships between quantities.

    b. Identify the constant of proportionality (unit rate) in tables, graphs, equations, diagrams, and verbal descriptions of proportional relationships.

    c. Represent proportional relationships by equations.

Student Activities Overview and Answer Key

Station 1

Students build, extend, and describe a geometric pattern made of toothpicks. Students are asked to work together to predict the number of toothpicks that would be needed to build the 10th stage of the pattern, and they are asked to write an expression that gives a general rule for the pattern.

Answers

1. 17, 21

2. 41

3. Possible answer: Multiply the number of the stage by 4 and add 1.

4. $4x + 1$

5. Possible explanation: To find the number of toothpicks, multiply the number of the stage by 4 and then add 1. So if the number of the stage is x, then the number of toothpicks is $4x + 1$.

Station 2

Students work together to analyze a pattern based on the cost of renting a car. They work together to complete a table of values. Then students write an expression that can be used to find the cost of renting a car for any number of days. They use the expression to find the cost of renting a car for 14 days and work backward (or write an equation) to find the number of days for which a car was rented if the total cost was $335.

Answers

Missing table values: 7 days, $195, $300

1. $20 + 35d$

2. The cost c is $20 plus $35 per day, or $20 + 35d$.

3. The cost is $20 + 35(14) = \$510$

4. 9 days

5. Find a value of d so that $20 + 35d = 335$. Use guess and check, or solve the equation algebraically.

Station 3

Students work together to make their own table based on a given situation about the number of buses needed to transport various numbers of students. Students write an expression for the pattern in the table and then use the expression to answer questions and make predictions about the real-world situation.

Answers

Sample table:

Number of buses (b)	1	2	3	4	5
Number of students (s)	24	48	72	96	120

1. $24b$

2. There are 24 students per bus, so the number of students for b buses is $24b$.

3. The number of students is $24(12) = 288$.

4. 9 buses. Find a value of b for which $24b = 216$ and find that $b = 9$.

Station 4

Students build, describe, and make predictions about a geometric pattern. To do so, they may use physical objects to model the pattern. Students work together to write expressions that describe various aspects of the geometric pattern.

Answers

1. 8 circular stickers and 25 square stickers

2. The number of circular stickers is equal to the number of the stage. The number of square stickers is 3 times number of the stage plus 1.

3. n

4. $3n + 1$

5. $4n + 1$

Materials List/Setup

Station 1   box of toothpicks

Station 2   none

Station 3   none

Station 4   small square tiles and small round tiles

Discussion Guide

To support students in reflecting on the activities and to gather some formative information about student learning, use the following prompts to facilitate a class discussion to "debrief" the station activities.

Prompts/Questions

1. What are some different tools, objects, or drawings that you can use to help you analyze a pattern?

2. What strategies can you use to help you translate information from words or tables into an algebraic expression?

3. Why is it useful to express relationships as algebraic expressions?

4. Is it ever possible for there to be more than one correct way to write an expression that represents a situation? Explain.

Think, Pair, Share

Have students jot down their own responses to questions, then discuss with a partner (who was not in their station group), and then discuss as a whole class.

Suggested Appropriate Responses

1. hands-on manipulatives, drawings of patterns, tables, etc.

2. Use verbal clues about operations. For example, the word "plus" means addition, which will be represented by a "+" symbol in the algebraic expression.

3. Algebraic expressions are more general than tables or graphs. You can use an expression to calculate additional values that are not shown in the table or graph.

4. Yes. For example, the expression $4s + 1$ can also be written as $1 + 4s$.

Possible Misunderstandings/Mistakes

- Incorrectly representing operations when writing an expression (e.g., writing $n + 2$ instead of $2n$)

- Multiplying the variable by an incorrect constant (e.g., writing $35 + 20d$ instead of $20 + 35d$)

- Incorrectly evaluating an expression for a specific value of the variable

Ratios and Proportional Relationships
Set 5: Representing Patterns and Relationships

Station 1

You will find some toothpicks at this station. Use them to help you with this activity.

Here are several stages of a pattern made from toothpicks.

Stage 1   **Stage 2**   **Stage 3**

Work with other students to build the pattern from toothpicks.

Work together to answer the questions below. When everyone agrees on an answer, write it on the line.

1. How many toothpicks do you need to make Stage 4 and Stage 5?

   _____

2. Predict the number of toothpicks you would need to make Stage 10.

   _____

3. Explain how you made this prediction.

   _____
   _____
   _____

4. Write an expression that gives the number of toothpicks that are needed at Stage x.

   _____

5. Explain how you came up with the expression.

   _____
   _____
   _____

Ratios and Proportional Relationships
Set 5: Representing Patterns and Relationships

Station 2

At this station, you will explore a pattern based on prices.

It costs $20 plus $35 per day to rent a car at Zippy Car Rentals.

Work with other students to complete the table.

Number of days (d)	1	3	5		8
Cost of car rental (c)	$55	$125		$265	

Work together to answer the questions below. When everyone agrees on an answer, write it on the line.

1. Write an expression that gives the cost of renting a car for d days.

   _____

2. Explain how you came up with the expression.

   _____

   _____

   _____

3. Show how you can use the expression to predict the cost of renting a car for 14 days.

   _____

   _____

   _____

4. Jerome rented a car at Zippy Car Rentals. The total cost was $335. For how many days did Jerome rent the car?

   _____

5. Explain how you found the answer to question 4.

   _____

   _____

   _____

Mathematics Station Activities for Common Core State Standards, Grade 7

Ratios and Proportional Relationships
Set 5: Representing Patterns and Relationships

Station 3

At this station, you will make your own table to help you explore a pattern.

A school bus can take 24 students on a field trip.

Work together to make a table that shows the number of buses and the corresponding number of students that can go on the field trip.

Work together to answer the questions below. When everyone agrees on an answer, write it on the line.

1. Write an expression that gives the number of student that can go on the field trip if there are *b* buses.

   _____

2. Explain how you came up with this expression.

   _____

   _____

3. Show how you can use the expression to find the number of students that can go on the trip if there are 12 buses.

   _____

   _____

4. How many buses are needed to take 216 students on the trip? Explain.

   _____

   _____

   _____

Ratios and Proportional Relationships
Set 5: Representing Patterns and Relationships

Station 4

At this station, you will find some tiles that you may use to help you analyze a pattern.

Tessa is using circular stickers and square stickers to make a pattern. Here are the first three stages of her pattern.

Stage 1　　　　**Stage 2**　　　　**Stage 3**

Work together to answer these questions. When everyone agrees on an answer, write it on the lines below.

1. How many circular stickers will Tessa need for Stage 8? _____

   How many square stickers will she need for Stage 8? _____

2. Explain how you found the answers to the questions above.

   _____

   _____

   _____

3. Write an expression that gives the number of circular stickers at stage *n*.

   _____

4. Write an expression that gives the number of square stickers at stage *n*.

   _____

5. Write an expression that gives the total number of stickers at stage *n*.

   _____

The Number System

Set 1: Adding and Subtracting Rational Numbers

Goal: To provide opportunities for students to develop concepts and skills related to addition and subtraction

Common Core State Standards

The Number System

Apply and extend previous understandings of operations with fractions to add, subtract, multiply, and divide rational numbers.

7.NS.1. Apply and extend previous understandings of addition and subtraction to add and subtract rational numbers; represent addition and subtraction on a horizontal or vertical number line diagram.

a. Describe situations in which opposite quantities combine to make 0.

b. Understand $p + q$ as the number located a distance $|q|$ from p, in the positive or negative direction depending on whether q is positive or negative. Show that a number and its opposite have a sum of 0 (are additive inverses). Interpret sums of rational numbers by describing real-world contexts.

c. Understand subtraction of rational numbers as adding the additive inverse, $p - q = p + (-q)$. Show that the distance between two rational numbers on the number line is the absolute value of their difference, and apply this principle in real-world contexts.

d. Apply properties of operations as strategies to add and subtract rational numbers.

Student Activities Overview and Answer Key

Station 1

Students model the sum of two integers using integer chips. They work as a group to ensure that the sum is modeled correctly. Then they use the model to find the sum of the integers.

Answers

1. –2
2. 6
3. –8
4. 7
5. 3
6. 0

Possible explanations: Count the remaining chips. If they are all yellow, the sum is the positive integer given by the number of remaining chips. If they are all red, the sum is the negative integer given by the number of remaining chips.

Station 2

Students are given a set of cards with integers written on them. They choose two cards at random and flip a coin to determine whether the two integers are to be added or subtracted. Students work together to find the sum or difference. When all students in the group agree on the answer, they write it down and repeat the process by choosing new integers to work with.

Answers

Answers will depend upon the numbers that are chosen.

Station 3

Students are given two sets of cards with rational numbers written on them. They choose one card from each set and work together to find the difference of the rational numbers. At the end of the activity, students discuss the strategies they used to subtract the numbers.

Answers

Answers will depend upon the cards that are chosen.

Possible strategies: If the cards show a decimal and a fraction, convert both numbers to decimals or both to fractions. If the number being subtracted is negative, change the problem to an addition problem. Model the subtraction problem with a number line.

Station 4

Students use a number cube to generate pairs of decimals. They work together to decide which of the two decimals in each pair is greater. Then they work together to subtract the smaller number from the larger one. All students in the group should agree on each answer.

Answers

Answers will depend upon the numbers that are rolled.

Possible strategies: Write the decimals so that one is above the other and so they are aligned on the decimal point. Add placeholder zeros if needed. Subtract the decimals as if they were whole numbers, placing the decimal in the answer in the same position as the decimals in the two given numbers.

Materials List/Setup

Station 1   integer chips

Station 2   a penny
set of 8 index cards with the following numbers written on them:
−12, −9, −7, −3, 4, 6, 11, 12

Station 3   10 index cards, prepared as follows:

- set of 5 cards, labeled "A" on the back with the following rational numbers written on the front:
  −$\frac{1}{2}$, 2, −3, 4.5, −6.5

- set of 5 cards, labeled "B" on the back with the following rational numbers written on the front:
  0.5, −2, 4, 5 $\frac{1}{2}$, −7

- Place the cards face-down at the station, so that only the backs (A or B) are visible.

Station 4   number cube (numbers 1–6)

Discussion Guide

To support students in reflecting on the activities and to gather some formative information about student learning, use the following prompts to facilitate a class discussion to "debrief" the station activities.

Prompts/Questions

1. How do you add two integers with the same sign?

2. How do you add two integers with opposite signs?

3. What can you say about the sum of two numbers that are opposites, such as 3.5 and −3.5?

4. How do you use paper and pencil to add or subtract two decimals?

Think, Pair, Share

Have students jot down their own responses to questions, then discuss with a partner (who was not in their station group), and then discuss as a whole class.

Suggested Appropriate Responses

1. Add the absolute values of the integers. Use the same sign for the answer as the sign of the given integers.

2. Find the absolute value of the integers. Subtract the smaller absolute value from the greater absolute value. Use the sign of the integer with the greater absolute value.

3. The sum of the numbers is zero.

4. Write the decimals so that one is above the other and so they are aligned on the decimal point. Add zeros as placeholders if needed. Add or subtract the decimals as if they were whole numbers, placing the decimal in the answer in the same position as the decimals in the two given numbers.

Possible Misunderstandings/Mistakes

*   Adding the absolute values of two integers when finding a sum of integers with opposite signs, rather than subtracting the absolute values (e.g., writing $-3 + 2 = -5$)

*   Using the wrong sign for the sum or difference of integers (e.g., writing $8 + (-6) = -2$)

*   Not recognizing that opposite integers always have a sum of zero

The Number System
Set 1: Adding and Subtracting Rational Numbers

Station 1

At this station, you will use integer chips to find the sum of two integers.

The yellow side of an integer chip shows +1. The red side shows –1. A pair of opposite chips (one yellow and one red) shows 0.

Work together using integer chips to show the first integer in each sum. Then use more integer chips to show the second integer in the sum. Work with other students to make sure each integer is shown correctly.

Then find the sum of the two integers by removing pairs of opposite chips and counting the chips that remain.

1. $5 + (-7) =$ _____

2. $-2 + 8 =$ _____

3. $-3 + (-5) =$ _____

4. $-1 + 8 =$ _____

5. $7 + (-4) =$ _____

6. $6 + (-6) =$ _____

After removing the opposite pairs, how did you decide what integer was shown? Explain your method.

_____

_____

_____

_____

_____

The Number System
Set 1: Adding and Subtracting Rational Numbers

Station 2

You will find a stack of cards and a penny at this station.

Choose two of the cards without looking. Turn the cards over and write the numbers in the boxes below. Then flip the penny. If it lands heads up, write "+" on the line between the numbers. If it lands tails up, write "−" on the line.

☐ ____ ☐

Work with other students to solve the addition or subtraction problem. When everyone agrees on the answer, write it below.

_____

Repeat the process three more times.

   Answer: _____

☐ ____ ☐   Answer: _____

   Answer: _____

The Number System

Set 1: Adding and Subtracting Rational Numbers

Station 3

You will be given two sets of cards labeled A and B. Choose one card labeled A and one card labeled B.

Turn the cards over. Subtract the rational number on card B from the rational number on card A. Write the subtraction problem below.

_____

Work as a group to subtract the numbers. Everyone in the group should agree on your answer. Write the answer below.

_____

Put the cards back. Mix up the cards. Repeat the process above three more times.

Subtraction problem: _____

Answer: _____

Subtraction problem: _____

Answer: _____

Subtraction problem: _____

Answer: _____

Explain the strategies you used to solve the subtraction problems.

_____

_____

_____

_____

_____

_____

_____

The Number System
Set 1: Adding and Subtracting Rational Numbers

Station 4

You will need a number cube for this activity. Use the number cube to create decimals.

Roll the number cube six times. Write the numbers in the six boxes below to create two decimals.

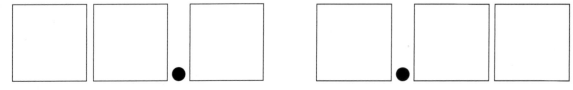

Work together to decide which of the two decimals is greater.

Subtract the smaller number from the larger one. Write the subtraction problem below.

_____

Work with other students to subtract the decimals. When everyone agrees on the answer, write it below.

_____

Repeat the process using the boxes below.

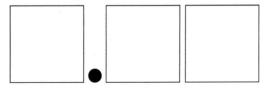 

Subtraction problem: _____

Answer: _____

Explain the strategies you used to subtract the decimals.

_____

_____

_____

_____

The Number System

Set 2: Multiplying and Dividing Rational Numbers

Goal: To provide students opportunities to develop concepts and skills related to multiplication and division

Common Core State Standards

The Number System

Apply and extend previous understandings of operations with fractions to add, subtract, multiply, and divide rational numbers.

    7.NS.2.   Apply and extend previous understandings of multiplication and division and of fractions to multiply and divide rational numbers.

        a.   Understand that multiplication is extended from fractions to rational numbers by requiring that operations continue to satisfy the properties of operations, particularly the distributive property, leading to products such as $(-1)(-1) = 1$ and the rules for multiplying signed numbers. Interpret products of rational numbers by describing real-world contexts.

        b.   Understand that integers can be divided, provided that the divisor is not zero, and every quotient of integers (with non-zero divisor) is a rational number. If p and q are integers, then $-(p/q) = (-p)/q = p/(-q)$. Interpret quotients of rational numbers by describing real-world contexts.

        c.   Apply properties of operations as strategies to multiply and divide rational numbers.

Student Activities Overview and Answer Key

Station 1

Students are given a set of cards with negative integers written on them. They choose two cards at random and flip a coin to determine whether the two integers are to be multiplied or divided. Students work together to find the product or quotient. When all students in the group agree on the answer, they write it down and repeat the process by choosing new integers to work with.

Answers

Answers will depend upon the numbers that are chosen.

Possible conclusion: All of the products and quotients are positive (i.e., the product or quotient of two negative numbers is positive).

Station 2

Students model the product of integers using integer chips. In each case, one integer in the product is positive and the other is negative. After working together to model a series of such products, students reflect on their work and write a statement about what they observe. In this way, students see that the product of a positive integer and a negative integer is always a negative integer.

Answers

$-3 \times 5 = -15$

1. -8 .   4. -7

2. -18 5. -16

3. -10 6. -6

Possible conclusion: The product of a positive integer and a negative integer is always a negative integer.

Station 3

Students are given two sets of cards with positive and negative rational numbers written on them. They choose one card from each set and work together to find the quotient of the rational numbers. At the end of the activity, students discuss the strategies they used to divide the rational numbers.

Answers

Answers will depend upon the cards that are chosen.

Possible strategies: First convert mixed numbers to improper fractions. Convert decimals to fractions. Divide the fractions by changing the operation to multiplication and changing the divisor to its reciprocal. Then multiply and simplify the answer. Place the correct sign (+ or –) in front of the quotient.

Station 4

Students are given a set of ten cards with integers or expressions written on them. Students work together to find pairs of cards that show the same value. When all the cards have been paired, students work together to check that integers or expressions on the pairs of cards are equal.

Answers

1. $-4 \times (-6) = 24$; $8 \div (-2) = -4$; $-16 \times 2 = 4 \times (-8)$; $-16 \div (-8) = -2 \times (-1)$; $6 \times (-4) = -48 \div 2$

2. Possible strategies: First sort cards into two groups depending on whether the value shown on each card is positive or negative. Calculate products and quotients using the rules of integer arithmetic.

Materials List/Setup

Station 1   a penny

set of 8 index cards with the following numbers written on them:

−1, −2, −4, −6, −8, −12, −16, −32

Station 2   integer chips

Station 3   10 index cards, prepared as follows:

- set of 5 cards, labeled "A" on the back with the following rational numbers written on the front:

  −8, −20, 6 $\frac{1}{2}$, 8.5, −3 $\frac{1}{2}$

- set of 5 cards, labeled "B" on the back with the following rational numbers written on the front:

  $\frac{1}{2}$, $\frac{1}{4}$, −0.5, −2, 4

- Place the cards face-down at the station, so that only the backs (A or B) are visible.

Station 4   10 index cards with the following numbers or expressions written on them:

−4 × (−6), 24, 8 ÷ (−2), −4, −16 × 2, 4 × (−8), −16 ÷ (−8), −2 × (−1), 6 × (−4), −48 ÷ 2

Discussion Guide

To support students in reflecting on the activities and to gather some formative information about student learning, use the following prompts to facilitate a class discussion to "debrief" the station activities.

Prompts/Questions

1. What can you say about the product or quotient of two integers with the same sign?

2. What can you say about the product or quotient of two integers with opposite signs?

3. How is multiplication of integers like multiplication of whole numbers? How is it different?

4. How are multiplication and division of signed numbers similar to each other?

Think, Pair, Share

Have students jot down their own responses to questions, then discuss with a partner (who was not in their station group), and then discuss as a whole class.

Suggested Appropriate Responses

1. The product or quotient is positive.

2. The product or quotient is negative.

3. To multiply two integers, multiply their absolute values, which are whole numbers. The step that is different from multiplication of whole numbers is the final step in which you write the correct sign (positive if both numbers have the same sign, negative if the numbers have opposite signs).

4. The rules for determining the sign of the answer are the same. If the two numbers being multiplied or divided have the same sign, the answer is positive. If the signs are different, the answer is negative.

Possible Misunderstandings/Mistakes

- Writing an incorrect sign when multiplying or dividing integers (e.g., writing $-3 \times (-4) = -12$)

- Incorrectly converting mixed numbers to fractions

- Incorrectly converting decimals to fractions or vice versa

The Number System
Set 2: Multiplying and Dividing Rational Numbers

Station 1

You will find a stack of cards and a penny at this station.

Choose two of the cards without looking. Turn the cards over and write the numbers in the boxes below. Then flip the penny. If it lands heads up, write a multiplication sign (×) on the line between the numbers. If it lands tails up, write a division sign (÷) on the line.

Work with other students to solve the multiplication or division problem. When everyone agrees on the answer, write it below.

_____

Repeat the process three more times.

Answer: _____

Answer: _____

Answer: _____

What do you notice about all the products and quotients that you found?

_____

_____

_____

The Number System
Set 2: Multiplying and Dividing Rational Numbers

Station 2

In this activity, you will use integer chips to help you multiply integers. The yellow side of an integer chip shows +1. The red side shows −1.

To multiply −3 × 5, first use integer chips to show the negative number, −3. There are 5 groups of −3, so use additional chips to show that you have 5 groups of −3.

Work together to decide what integer is shown by all the chips. This is the product. Write it below.

_____

Work with other students using integer chips to find each product.

1. $-2 \times 4 = $ _____

2. $3 \times (-6) = $ _____

3. $-5 \times 2 = $ _____

4. $7 \times (-1) = $ _____

5. $-4 \times 4 = $ _____

6. $-3 \times 2 = $ _____

Look at all the products you found. What can you say about the product of a positive integer and a negative integer?

_____

_____

_____

_____

_____

The Number System
Set 2: Multiplying and Dividing Rational Numbers

Station 3

At this station, you will find two sets of cards labeled A and B. Choose one card labeled A and one card labeled B.

Turn the cards over. Divide the rational number on card A by the rational number on card B. Write the division problem below.

_____

Work as a group to divide the rational numbers. Everyone in the group should agree on your answer. Write the answer below.

_____

Put the cards back. Mix up the cards. Repeat the above process three more times.

Division problem: _____

Answer: _____

Division problem: _____

Answer: _____

Division problem: _____

Answer: _____

Explain at least three strategies that you could use to solve the division problems.

_____

_____

_____

_____

_____

The Number System
Set 2: Multiplying and Dividing Rational Numbers

Station 4

At this station, you will find cards with the following numbers or expressions written on them:

$$-4 \times (-6) \qquad 8 \div (-2) \qquad 6 \times (-4) \qquad 24 \qquad -16 \times 2$$

$$-16 \div (-8) \qquad -4 \qquad 4 \times (-8) \qquad -2 \times (-1) \qquad -48 \div 2$$

Work with other students to sort the cards into pairs. The values on the two cards in each pair should be equal. When you have sorted the cards into pairs, work together to check that the values on the two cards in each pair are equal.

1. Write the five pairs below.

   _____

   _____

   _____

   _____

   _____

2. Describe the strategies you used to sort the cards into pairs.

   _____

   _____

   _____

   _____

   _____

Expressions and Equations

Instruction

Goal: To provide students the opportunity to apply problem-solving skills and knowledge of rational numbers in any form, properties of operations, and converting between forms

Common Core State Standard

Solve real-life and mathematical problems using numerical and algebraic expressions and equations.

7.EE.3. Solve multi-step real-life and mathematical problems posed with positive and negative rational numbers in any form (whole numbers, fractions, and decimals), using tools strategically. Apply properties of operations to calculate with numbers in any form; convert between forms as appropriate; and assess the reasonableness of answers using mental computation and estimation strategies.

Student Activities Overview and Answer Key

Station 1

Students will work together to calculate a pay increase for a burger joint employee, including hourly increase, new hourly wage, expected earnings for a 12-hour week, and difference in earnings before and after the raise.

Answers

1. 78 cents

2. $8.58

3. $102.96

4. $9.36

Station 2

Students will work together to determine how much of a raise or how many additional hours of babysitting would be necessary to earn a certain amount of money in a certain amount of time.

Answers

1. $195

2. $55

3. 8.5 to 9 hours

4. He would need to charge $1.83 more per hour.

5. Answers will vary, but should add up to at least $250.

Station 3

Students will work together to determine how to place two posters on a bedroom wall, with even spacing on either side and between the two posters.

Answers

1. 50 inches

2. inches; 104 inches

3. 18 inches on either side and between

4. Answers will vary, but must add up to a total of 104 inches.

5. Diagrams should show two 25-inch wide posters spaced 18 inches apart with 18 inches of space on either side on a 104-inch-wide wall (8 ⅔ feet). Check diagram labels for accuracy.

Station 4

Students will work together to figure out what size containers to use to store lemonade. They will need to calculate the number of cups of lemonade needed for a party and what combination of containers will hold the lemonade.

Answers

1. 14 cups

2. 3.5 quarts

3. four 1-quart pitchers

4. 0.875 gallons

5. 1 gallon pitcher

6. Answers will vary, but should account for 14 cups of lemonade and should provide an explanation. For example, "One gallon pitcher is a better choice because then you only need one." OR "Four quart pitchers are better because they are easier to pour and everyone will have one nearby."

Materials List/Setup

Station 1   *optional:* calculators

Station 2   *optional:* calculators

Station 3   *optional:* calculators; tape measure

Station 4   *optional:* calculators; water; separate containers to hold 1 cup, 1 quart, and 1 gallon of water, respectively

Discussion Guide

To support students in reflecting on the activities and to gather some formative information about student learning, use the following prompts to facilitate a class discussion to "debrief" the station activities.

Prompts/Questions

1. How do you calculate a raise or other increase (in price, temperature, etc.)? What are the steps?

2. How do you work backward from an amount of money that you want or need in order to figure out how to earn/save it?

3. How do you convert between measurements such as from inches to feet, feet to yards, cups to quarts, quarts to gallons?

4. In a multi-step problem, how do you know if your answer is correct?

Think, Pair, Share

Have students jot down their own responses to questions, then discuss with a partner (who was not in their station group), and then discuss as a whole class.

Suggested Appropriate Responses

1. Multiply by the percent or fraction increase and then add to the original rate/price/ temperature.

2. Divide the total amount by the hourly/daily/weekly rate. If appropriate, first subtract the existing amount.

3. Remember, or look up the conversions—12 inches per foot and 3 feet per yard; 4 cups per quart and 4 quarts per gallon.

4. Substitute your solution back into the problem to see if it works. Do a "common sense" check to make sure it's reasonable.

Possible Misunderstandings/Mistakes

*   Difficulty converting a fraction or percent to value in dollars and cents

*   Forgetting one or more steps in a multi-step problem; e.g., failing to add or subtract from the original amount

*   Making errors in converting amongst units of measure, or using incorrect equivalents

*   Not using the scenario provided to help reason through the mathematics

Expressions and Equations
Set 1: Multi-Step Real-Life Problems with Rational Numbers

Station 1

Jayda works at a burger restaurant. She gets paid $7.80 per hour. Now that she's worked part-time for six months, she is getting a 10-percent raise!

Use the information in the problem to answer each question. Show and explain your math.

1. What is a 10-percent raise on $7.80 per hour?

2. How much will Jayda make per hour after her raise?

3. How much will Jayda earn in a week if she works for 12 hours at her new wage?

4. How much more will she be making this week than she did last week (if both are 12-hour work weeks)?

Expressions and Equations

Set 1: Multi-Step Real-Life Problems with Rational Numbers

Station 2

Damian plans to do a lot of babysitting this summer. He charges $6.50 per hour and usually babysits for 3 hours. He has agreed to babysit every Saturday from mid-June through August (10 times) for his neighbors.

Damian's parents said that they would split the cost of a new computer with him. It costs $500 and he has to save half that amount before school starts.

Use the information in the problem to answer each question. Show and explain your math.

1. How much money will Damian earn from the babysitting that he already has scheduled?

2. How much more money will he need to make a total of $250?

3. How many more hours will he have to babysit to earn the extra money that he needs?

4. How much more would Damian have to charge per hour if he wanted to make the whole $250, in the number of hours he already has scheduled to babysit?

5. If Damian wants to do a combination of raising his prices and working more hours, would that work? How could he do it and raise the $250 he needs?

Expressions and Equations

Set 1: Multi-Step Real-Life Problems with Rational Numbers

Station 3

Mike's brother is going to college, so now he has a bedroom to himself. Mike wants to put up 2 new movie posters on the wall that used to be his brother's. The wall is $8\frac{2}{3}$ feet wide and each poster is 25 inches wide. He wants the posters spaced out so there is the same amount of space on either side of the posters, and between them, too.

Use the information in the problem to answer each question. Show and explain your math. Then, make a diagram of the wall and the 2 posters and label the measurements.

1. How wide are the 2 posters together?

2. What unit of measurement will you have to use for the width of the wall? How wide is the wall in this unit of measurement?

3. How much space will there be between the posters and on either side, if they are spaced evenly?

continued

Expressions and Equations

Set 1: Multi-Step Real-Life Problems with Rational Numbers

4. How else could you space the posters on the wall? List the distances on either side and between them.

5. Draw a diagram of the wall showing how Mike arranged the posters. Label the measurements for the posters and the space on either side and between them.

Expressions and Equations

Set 1: Multi-Step Real-Life Problems with Rational Numbers

Station 4

Holly and Aubrey are having a party. They plan to serve lemonade. Including the two girls, there will be 14 guests at the party. Holly and Aubrey are assuming that each person will have a cup of lemonade.

Use the information in the problem to answer each question. You may also use cups, pitchers, and water if your teacher provides them. Show and explain your math.

1. How many cups of lemonade do Holly and Aubrey need for their party?

2. How many quarts of lemonade is that?

3. How many quart pitchers would Holly and Aubrey need to store and serve the lemonade?

4. How many gallons of lemonade do the number of cups equal?

5. How many gallon pitchers would Holly and Aubrey need to store and serve the lemonade?

6. Which is the better size pitcher to use—quart or gallon? Explain your reasoning.

Expressions and Equations

Set 2: Using Variables to Construct Equations and Inequalities

Goal: To provide students opportunities to interpret word problems and to construct simple equations and expressions to solve them

Common Core State Standards

Solve real-life and mathematical problems using numerical and algebraic expressions and equations.

7.EE.4. Use variables to represent quantities in a real-world or mathematical problem, and construct simple equations and inequalities to solve problems by reasoning about the quantities.

a. Solve word problems leading to equations of the form $px + q = r$ and $p(x + q) = r$, where p, q, and r are specific rational numbers. Solve equations of these forms fluently. Compare an algebraic solution to an arithmetic solution, identifying the sequence of the operations used in each approach.

b. Solve word problems leading to inequalities of the form $px + q > r$ or $px + q < r$, where p, q, and r are specific rational numbers. Graph the solution set of the inequality and interpret it in the context of the problem.

Student Activities Overview and Answer Key

Station 1

Students will write an equation, using variables, to represent the length, width, and perimeter of a dance floor. They will solve for width, given side length and perimeter. They will then write an equation for the area of the dance floor and solve using the information that they already have.

Answers

1. $2l + 2w = p$

2. $2(20) + 2w = 60$

    $40 + 2w = 60$

    $2w = 20$

    $w = 10$

3. $l \bullet w = a$ or $20 \bullet 10 = a$

4. $20 \bullet 10 = 200 \ m^2$

Station 2

Students will write an inequality, using variables, to represent the number of times Mia will have to watch her little brother, in addition to her regular allowance, to save $25 this week.

Answers

1. $\$10 + \$3x \geq \$25$

2. $x \geq 5$; Mia must watch her brother for 5 or more hours.

Station 3

Students will write two different forms of an equation, using variables, to represent the total number of lemonade sales two boys made in order to make $2.40. Students will then list at least three combinations that add up to the necessary total.

Answers

1. $\$0.50R + \$0.50A = T$, or $\$0.50(R + A) = T$

2. 48 cups of lemonade

3. Answers will vary, but should sum to 48.

Station 4

Students will write an inequality, using variables, to represent the number of magazine subscriptions they need to sell in order to reach or exceed their goal. Then they describe the solution to the inequality.

Answers

1. $\$150 + \$5x \geq \$500$

2. $x \geq 70$; they have to sell at least 70 more subscriptions.

Materials List/Setup

Station 1	calculators and scrap/scratch paper
Station 2	calculators and scrap/scratch paper
Station 3	calculators and scrap/scratch paper
Station 4	calculators and scrap/scratch paper

Discussion Guide

To support students in reflecting on the activities and to gather some formative information about student learning, use the following prompts to facilitate a class discussion to "debrief" the station activities.

Prompts/Questions

1. What is a variable?

2. How do you decide what values need to be represented by a variable?

3. What is the difference between an equation and an inequality?

4. Why is it important to label answers?

Think, Pair, Share

Have students jot down their own responses to questions, then discuss with a partner (who was not in their station group), and then discuss as a whole class.

Suggested Appropriate Responses

1. A variable is a letter used to represent an unknown value.

2. When writing an equation or inequality, you need to use a variable to represent each value that you don't know.

3. An equation has a finite number of answers. An inequality has a range (infinite number) of correct answers.

4. Units or labels make it clear what the answer actually represents. (Dollars? Inches? Subscriptions?)

Possible Misunderstandings/Mistakes

- Difficulty identifying which values need to be represented with a variable

- Difficulty selecting/assigning a variable (Point out that there is no "rule;" x, y, z are common; and using initials such as s for subscriptions, is helpful.)

- Lacking knowledge of standard formulas for perimeter, area, etc.

- Confusion about the difference between an equation and an inequality

- Confusion about $\leq$ versus $\geq$ in a given situation

- Failing to label answers

- Failing to understand what solutions represent vis-à-vis the problem

Expressions and Equations

Set 2: Using Variables to Construct Equations and Inequalities

Station 1

A dance floor has a perimeter of 60 meters. The dance floor is 20 meters long. Use this information to help you follow the directions below.

1. Write an equation using variables to represent the perimeter of the dance floor.

2. Solve the equation to find the dimensions of the dance floor.

3. Write an equation using variables to represent the area of the dance floor.

4. Solve the equation to find the area of the dance floor.

Expressions and Equations
Set 2: Using Variables to Construct Equations and Inequalities

Station 2

Mia's mother gives her $10 per week for allowance. Mia gets an additional $3 for every hour she spends watching her little brother. Mia wants to make at least $25 this week to spend shopping with her friends. Use this information to help you follow the directions below.

1. Write an inequality, using variables, to represent the number of times that Mia has to watch her brother in order to get at least $25.

2. Solve your inequality and describe the solution.

Expressions and Equations

Set 2: Using Variables to Construct Equations and Inequalities

Station 3

Raf and Ali set up two lemonade stands in their neighborhood. They each charge $0.50 for a cup of lemonade. At the end of the day, they meet to combine their earnings. Use this information to help you follow the directions below.

1. Write two different equations, using variables, to represent the amount of money that the boys will make.

2. Together, Raf and Ali got $24.00. How many cups did they sell?

3. Show at least three different combinations of lemonade sales for Raf and Ali that would add up to earnings of $24.00.

Expressions and Equations
Set 2: Using Variables to Construct Equations and Inequalities

Station 4

Ms. Mohammed's class is selling magazine subscriptions to earn money for a field trip. For every subscription sold, the class makes $5. They have already earned $150, but they need $500 total. Use this information to help you follow the directions and answer the questions below.

1. Write an inequality, using variables, to represent the number of subscriptions that will have to be sold to make at least $500.

2. Solve the inequality and describe the solution. How many more subscriptions does Ms. Mohammed's class have to sell?

Expressions and Equations

Goal: To provide opportunities for students to develop concepts and skills related to solving inequalities

Common Core State Standard

Expressions and Equations

Solve real-life and mathematical problems using numerical and algebraic expressions and equations.

7.EE.4. Use variables to represent quantities in a real-world or mathematical problem, and construct simple equations and inequalities to solve problems by reasoning about the quantities.

b. Solve word problems leading to inequalities of the form $px + q > r$ or $px + q < r$, where p, q, and r are specific rational numbers. Graph the solution set of the inequality and interpret it in the context of the problem.

Student Activities Overview and Answer Key

Station 1

Students are given a series of inequalities and a number cube. For each inequality, they roll the number cube and then work together to decide if the number shown on the cube is a solution of the inequality. Students explain the strategies they used to decide whether each value was a solution.

Answers

1–3. Answers will depend upon numbers rolled. 4. yes; 5. no

Possible strategies: Substitute the value for the variable. Simplify and check to see if the resulting inequality is true.

Station 2

In this activity, students work together to use number lines to help them solve inequalities. To do so, they test various values of the variable in the given inequalities, and check to see whether each value is a solution. They keep track of the values that are solutions by marking them on a number line. After testing enough values to see a pattern, students shade the values that represent all solutions of the inequality. Then they write the solution algebraically.

Answers

1. $x < 2$

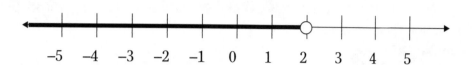

2. $x > -2$

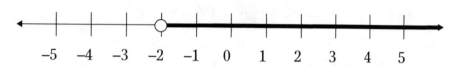

3. $x < 1$

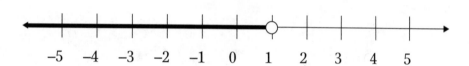

4. $x < 1$

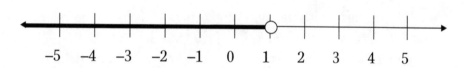

5. $x < 2$

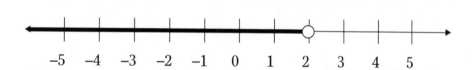

Station 3

In this activity, students work together to match a set of given inequalities with a set of given solutions. Once students have paired each inequality with its correct solution, they discuss the strategies they used to solve the problem.

Answers

The cards should be paired as follows: $5x + 2 < 12$ and $x < 2$; $4x + 3 < -5$ and $x < -2$; $-3x < 6$ and $x > -2$; $-x/_2 > 2$ and $x < -4$; $3x + 1 > -11$ and $x < -4$; $x/_4 + 1 > 2$ and $x > 4$.

Possible strategies: Solve each inequality using inverse operations and look for the solution among the given choices.

Station 4

Students are given a set of inequalities and a set of real-world situations. They work together to match each situation to an inequality. Then they solve the inequality. At the end of the activity, students explain the strategies they used to match the inequalities to the situations.

Answers

1. $10x + 5 < 105$, $x < 11$; 2. $5x + 10 < 105$, $x < 19$; 3. $10x - 5 > 105$, $x > 11$; 4. $5(x - 10) < 105$, $x < 31$

Possible strategies: Use the words or phrases that refer to arithmetic operations as clues to identifying the corresponding inequalities. For example, a decrease corresponds to subtraction. Match words to inequalities. For example, "no more than but not including" refers to a "less than" inequality (<).

Materials List/Setup

Station 1   number cube (numbers 1–6)

Station 2   none

Station 3   set of index cards with the following inequalities written on them:
$5x + 2 < 12$, $4x + 3 < -5$, $-3x < 6$, $-\frac{x}{2} > 2$, $3x + 1 > -11$, $\frac{x}{4} + 1 > 2$
set of index cards with the following solutions written on them:
$x < -4$, $x > -4$, $x < -2$, $x > -2$, $x < 2$, $x > 4$

The two sets of cards should be placed in two piles, face-up, on a table or desk at the station.

Station 4   none

Discussion Guide

To support students in reflecting on the activities and to gather some formative information about student learning, use the following prompts to facilitate a class discussion to "debrief" the station activities.

Prompts/Questions

1. What is the difference between < and >?

2. How do you check to see if a value of the variable is a solution of an inequality?

3. How is the solution of an inequality different from the solution of an equation?

4. How do you solve an equality using algebra?

Think, Pair, Share

Have students jot down their own responses to questions, then discuss with a partner (who was not in their station group), and then discuss as a whole class.

Suggested Appropriate Responses

1. The symbol < means "less than." The symbol > means "more than."

2. Substitute the value for the variable in the inequality. Check to see if the resulting inequality is true. If it is, the value is a solution.

3. In general, the solution of an inequality is itself an inequality (a range of values). The solution of an equation is usually a single value (or several discrete values).

4. Use inverse operations, as when solving an equation, to isolate the variable on one side of the inequality. If you multiply or divide by a negative number, reverse the direction of the inequality.

Possible Misunderstandings/Mistakes

- Using an incorrect operation to solve an inequality (e.g., solving $x + 2 < 5$ by adding 2 to both sides)

- Incorrectly translating verbal expressions to inequalities (e.g., representing the phrase "less than" by > rather than <)

- Forgetting to reverse the direction of the inequality when multiplying or dividing by a negative number

Expressions and Equations

Set 3: Solving Inequalities

Station 1

You will find a number cube at this station.

For each inequality, roll the number cube and write the number in the box. Then work together to decide if this value of the variable is a solution of the inequality. Write "yes" or "no" on the line provided.

1. $2x + 1 < 7$ Solution? _____

2. $3x - 4 > 5$ Solution? _____

3. $-3x < -12$ Solution? _____

4. $\dfrac{x}{2} + 1 < 5$ Solution? _____

5. $1 - x > 0$ Solution? _____

6. Explain the strategies you used to decide whether each value was a solution of the inequality.

   _____

   _____

   _____

Expressions and Equations
Set 3: Solving Inequalities

Station 2

You can use number lines to help you solve inequalities.

For each inequality, work together to test different values of the variable to see if they are solutions of the inequality. If a value is a solution, draw a solid dot at that value on the number line. Test at least five different values for each inequality.

When you think you know what the solution set of an inequality looks like, shade the correct part of the number line to show all the solutions.

Finally, write the solution in the space provided.

1. $2x - 3 < 1$

   Solution: _____

2. $3x + 1 > -5$

   Solution: _____

3. $-2x < -2$

   Solution: _____

4. $4x - 2 < 2$

   Solution: _____

5. $\dfrac{x}{2} + 3 < 4$

   Solution: _____

Expressions and Equations
Set 3: Solving Inequalities

Station 3

At this station, you will work with other students to match inequalities to their solutions.

You will find a set of cards with the following inequalities written on them:

$$5x + 2 < 12 \qquad 4x + 3 < -5 \qquad -3x < 6 \qquad -\frac{x}{2} > 2 \qquad 3x + 1 > -11 \qquad \frac{x}{4} + 1 > 2$$

You will also find a set of cards with these solutions written on them:

$$x < -4 \qquad x > -4 \qquad x < -2 \qquad x > -2 \qquad x < 2 \qquad x > 4$$

Work together to match each inequality with its solution. When everyone agrees on the answers, write the matching pairs below.

_____

_____

_____

_____

_____

_____

Explain the strategies you used to match up the cards.

_____

_____

_____

Expressions and Equations
Set 3: Solving Inequalities

Station 4

At this station, you will match inequalities to real-world situations and then solve the inequalities.

Work with other students to match each situation to one of the following inequalities. When everyone agrees on the correct inequality, write it on the line provided. Then work together to solve it.

$$5x + 10 < 105 \qquad 10x + 5 < 105 \qquad 5(x - 10) < 105 \qquad 10x - 5 > 105$$

1. Mai rents DVDs by mail. There is a one-time membership fee of $5 and the service costs $10 per month. Mai wants to spend less than $105. For how many months can she rent DVDs with this service?

   Inequality: _____

   Solution: _____

2. Tyrone bought 5 trays of food for a party. The price of each tray of food was the same. He also spent $10 on paper plates, napkins, and utensils. Altogether, he spent less than $105. What was the price of each tray of food?

   Inequality: _____

   Solution: _____

3. Mr. Garcia ordered 10 copies of a novel for students in his English class. He had a coupon for $5 off the total price of the order. The total cost of the order, before tax, came to more than $105. What was the price of each novel?

   Inequality: _____

   Solution: _____

continued

Expressions and Equations
Set 3: Solving Inequalities

4. Rachel bought 5 pairs of jeans. Each pair of jeans had the same price. She had a coupon for $10 off the price of each pair of jeans. The total cost of the jeans, before tax, came to less than $105. What was the price of each pair of jeans?

Inequality: _____

Solution: _____

Explain the strategies you used to match the inequalities to the situations.

_____

_____

_____

Geometry

Set 1: Similarity and Scale

Goal: To provide opportunities for students to develop concepts and skills related to the properties of similarity

Common Core State Standards

Geometry

Draw, construct, and describe geometrical figures and describe the relationships between them.

7.G.1. Solve problems involving scale drawings of geometric figures, including computing actual lengths and areas from a scale drawing and reproducing a scale drawing at a different scale.

7.G.2. Draw (freehand, with ruler and protractor, and with technology) geometric shapes with given conditions. Focus on constructing triangles from three measures of angles or sides, noticing when the conditions determine a unique triangle, more than one triangle, or no triangle.

Student Activities Overview and Answer Key

Station 1

At this station, students draw different triangles with the same three angles. They then measure their sides and compare the ratios of the sides. Students reflect on their findings.

Answers: The ratios are approximately the same for all the triangles; they are similar; the ratio of the sides in similar triangles is the same

Station 2

Students imagine they are architects. They use a scale diagram of a room to determine the actual measurements of a room.

Answers: They are similar; it is useful to share plans with other people; they cannot possibly make their plans the same size as what they are building, etc.

Station 3

Students look at a model car. They take measurements and use scale factor to determine the measurements of the actual car. They then reflect on the relationship between the model car and the actual car.

Answers: All measurements vary; similar because all the measurements of the model car are proportional to the actual car, but not exact

Station 4

Students explore the relationship between similarity and area in triangles. They construct their own triangles and find the ratio of the areas. They then reflect on their observations.

Answers: 6 sq in.; 24 sq in.; 4 to 1; answers will vary; answers will vary; 4 to 1; the ratio is 4 to 1; the base and height are both increased by a factor of 2, and 2×2 is 4

Materials List/Setup

Station 1	enough pencils, protractors, rulers, and calculators for all group members
Station 2	enough rulers and calculators for all group members
Station 3	a model car with scale factor; enough rulers and calculators for all group members
Station 4	enough rulers and calculators for all group members

Discussion Guide

To support students in reflecting on the activities and to gather some formative information about student learning, use the following prompts to facilitate a class discussion to "debrief" the station activities.

Prompts/Questions

1. How are similarity and congruence different?

2. When is scale factor important in real life?

3. What is the ratio of a large rectangle's area to a small rectangle's area if the dimensions are increased by a factor of 2?

4. If you had a figure and wanted to construct a similar figure, what information would be important to know?

Think, Pair, Share

Have students jot down their own responses to questions, then discuss with a partner (who was not in their station group), and then discuss as a whole class.

Suggested Appropriate Responses

1. Similar figures have the same angles. Congruent figures have the same angles and sides.

2. many possibilities—in blueprints

3. 4 to 1

4. the angles, proportions of the sides

Possible Misunderstandings/Mistakes

- Trouble accurately measuring angles—if incorrect, data and lesson will be incorrect
- Trouble accurately using a ruler to measure sides—if incorrect, data and lesson will be incorrect
- Confusing units when dealing with scale factor

Geometry
Set 1: Similarity and Scale

Station 1

At this station, you will find rulers, pencils, calculators, and protractors. You will be using these materials to draw a triangle and make observations.

Each group member should draw a triangle in the space below. The angles of the triangle should be 82°, 28°, and 70°. The 82° angle is angle A. The 28° angle is angle B. And the 70° angle is angle C.

Each person should measure the sides of his/her triangle. Compile your data in the table below.

Group member	Length AB (cm)	Length BC (cm)	Length AC (cm)	AB/BC	BC/AC	AB/AC

continued

Geometry
Set 1: Similarity and Scale

What do you notice about the ratio of the sides? _____

_____

_____

What types of triangles are these? What is the relationship between any two of the triangles?

_____

_____

What do you notice about similar triangles? _____

_____

_____

Geometry
Set 1: Similarity and Scale

Station 2

At this station, you will take on the role of an architect. You will find rulers and calculators to help you with this role.

Below is a diagram of a room in a house. Work as a group to figure out the dimensions of the actual room in the house. For every inch in the diagram, the real room is 10 feet.

Use your ruler to find the dimensions of the diagram. Put the actual lengths on the slightly larger diagram.

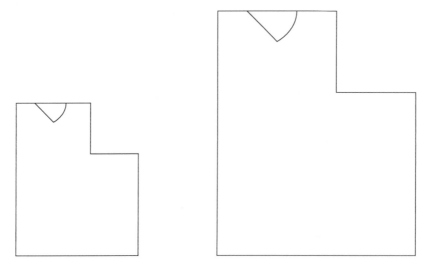

What is the relationship between the diagram and the actual room? _____

_____

Why is it useful for architects to make models? _____

_____

_____

_____

Geometry
Set 1: Similarity and Scale

Station 3

At this station, you will find a model car and the scale factor that describes the relationship between the size of the model and the real car.

Draw a diagram of this car in the space below without the measurements.

Use the scale factor to set up appropriate proportions. Show your work below.

Write the measurements of the real car on your drawing.

Discuss with your group. What word describes the relationship between the model car and the real car—congruent or similar? Explain.

_____

_____

_____

Geometry
Set 1: Similarity and Scale

Station 4

At this station, you will find a ruler and a calculator. You will use these to look at the relationship between the areas of similar triangles.

Look at the triangles below.

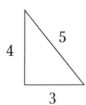
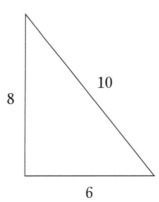

What is the area of the smaller triangle? _____

What is the area of the larger triangle? _____

What is the ratio of area of the larger triangle to the smaller triangle? _____

Draw your own triangle. Measure the height and the base. Create a triangle with sides that are twice as long. Find the area of the two triangles.

Area of small triangle: _____

Area of large triangle: _____

Ratio of area of large triangle to area of small triangle: _____

Discuss with your group. What is true about the area of similar triangles when the sides are increased by a factor of two?

_____

_____

Why do you think this is true? _____

_____

_____

Geometry

Goal: To provide opportunities for students to develop concepts and skills related to sketching, modeling, and describing the cross-sections of three-dimensional figures

Common Core State Standard

Geometry

Draw, construct, and describe geometrical figures and describe the relationships between them.

7.G.3. Describe the two-dimensional figures that result from slicing three-dimensional figures, as in plane sections of right rectangular prisms and right rectangular pyramids.

Student Activities Overview and Answer Key

Station 1

Students use clay and mold it into a cone. They then cut the cone in various directions and observe how the different cuts produce different cross-sections.

Answers: circle; parabola or part of an ellipse; an ellipse; the cross-sections are different depending on what way we cut the cone

Station 2

Students explore the cross-sections of a rectangular prism. They draw cross-sections from three different cuts. They then notice the similarities between the cuts.

Answers: rectangle; rectangle; rectangle; they are all rectangles

Station 3

Students have three cross-sections of a solid, and use this information to determine what solid the cross-sections are from. They then explain their strategy for successfully completing this task.

Answers: a pyramid; answers will vary

Station 4

Students use three marshmallows to investigate the cross-sections of a cylinder. They draw each cross-section, and then compare and contrast the results.

Answers: Answers will vary.

Materials List/Setup

Station 1   clay; plastic knives

Station 2   a rectangular prism

Station 3   none

Station 4   3 marshmallows; a plastic knife

Discussion Guide

To support students in reflecting on the activities and to gather some formative information about student learning, use the following prompts to facilitate a class discussion to "debrief" the station activities.

Prompts/Questions

1. What is an example of when you see the cross-section of a rectangular prism?

2. How would you get a point as a cross-section in a cone?

3. What are two figures that have the same shaped cross-sections (not necessarily the same size) if you cut them horizontally, vertically, and diagonally?

4. What are two figures that have different shaped cross-sections if you cut them horizontally, vertically, and diagonally?

Think, Pair, Share

Have students jot down their own responses to questions, then discuss with a partner (who was not in their station group), and then discuss as a whole class.

Suggested Appropriate Responses

1. many examples—at the deli when they cut cheese

2. make a slice at the very tip of the point of the cone

3. sphere, cube, rectangular prism

4. many examples—cylinder, pyramid, cone, etc.

Possible Misunderstandings/Mistakes

- Having trouble visualizing cross-sections

- Having trouble cutting the clay or marshmallows

- Confusing any of the three different cross-sections

Geometry
Set 2: Sketching, Modeling, and Describing 3-D Figures

Station 1

At this station, you will find clay and plastic knives.

Mold your clay into a cone. Make a horizontal cut. What shape do you get when you look down at the cut? Draw what you see.

Mold your clay back into a cone. Make a vertical cut. What shape do you get when you look down at the cut? Draw what you see.

Mold your clay into a cone for the third time. Make a diagonal cut. What shape do you get when you look down at the cut? Draw what you see.

How does the direction of the cross-section of a cone change what it looks like? Discuss with your group.

_____

_____

_____

Geometry
Set 2: Sketching, Modeling, and Describing 3-D Figures

Station 2

At this station, you will find a rectangular prism which will help you visualize different cross-sections. Discuss and answer the question below.

If you made a horizontal cut, what would the cross-section look like? Draw it below.

If you made a vertical cut, what would the cross-section look like? Draw it below.

If you made a diagonal cut, what would the cross-section look like? Draw it below.

What do you notice about all three cross-sections? _____

_____

_____

Geometry
Set 2: Sketching, Modeling, and Describing 3-D Figures

Station 3

At this station, you will be looking at a set of cross-sections and using them to figure out what solid they are from.

Look at the cross-sections below. They are horizontal, vertical, and diagonal, in that order.

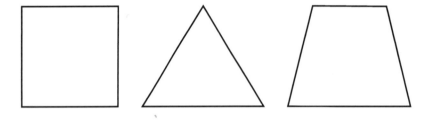

What solid is each cross-section from? Discuss with your group.

_____

_____

_____

Explain your strategy for figuring that out. _____

_____

_____

_____

A cone is similar to a pyramid, except with a circle base. Draw the cross-sections of a cone below.

Geometry

Set 2: Sketching, Modeling, and Describing 3-D Figures

Station 4

At this station, you will find three marshmallows and a plastic knife. You will use these to investigate the cross-sections of a cylinder.

Cut the first marshmallow horizontally. What does it look like? Draw the cross-section below.

Cut the second marshmallow vertically. What does it look like? Draw the cross-section below.

Cut the third marshmallow diagonally. What does it look like? Draw the cross-section below.

Discuss with your group. Describe the similarities and differences in the cross sections.

_____

_____

_____

_____

Statistics and Probability

Set 1: Collecting, Organizing, and Analyzing Data

Goal: To provide opportunities for students to develop concepts and skills related to data collection and analysis

Common Core State Standards

Statistics and Probability

Investigate chance processes and develop, use, and evaluate probability models.

7.SP.5. Understand that the probability of a chance event is a number between 0 and 1 that expresses the likelihood of the event occurring. Larger numbers indicate greater likelihood. A probability near 0 indicates an unlikely event, a probability around 1/2 indicates an event that is neither unlikely nor likely, and a probability near 1 indicates a likely event.

7.SP.6. Approximate the probability of a chance event by collecting data on the chance process that produces it and observing its long-run relative frequency, and predict the approximate relative frequency given the probability.

7.SP.7. Develop a probability model and use it to find probabilities of events. Compare probabilities from a model to observed frequencies; if the agreement is not good, explain possible sources of the discrepancy.

b. Develop a probability model (which may not be uniform) by observing frequencies in data generated from a chance process.

7.SP.8. Find probabilities of compound events using organized lists, tables, tree diagrams, and simulation.

Student Activities Overview and Answer Key

Station 1

Students will be generating their own data and turning it into a scatter plot. They will examine and answer questions pertaining to this scatter plot.

Answers: answers may vary; answers may vary

Station 2

Students will formulate different questions that they would be able to go out and collect data for, remembering that it is important to collect enough data to draw valid conclusions. They then choose one question to look at more closely. Students reflect on how they would work towards answering that question, and why it is important to collect a lot of data.

Answers: answers may vary; answers may vary; answers may vary; answers may vary; If we do not collect enough data, we will not be able to see true trends in the data.

Station 3

Students try to answer the question of which side of a number cube comes up most. They conduct an experiment to help answer this question, and then develop other questions that could be answered by their data.

Answers: answers may vary; answers may vary; answers may vary, for example, Which number comes up least? Do all numbers come up about the same number of times?

Station 4

Students roll a number cube and find the mean at different intervals. They then compare the means and determine which is closest to the expected value. This helps point out why it is important to have a large sample size (30 or more).

Answers: answers will vary; answers will vary; answers will vary; the mean of 30 rolls should be closest to the expected value of a roll; the mean of 30 rolls should be closest to 3.5 because there was an opportunity to account for flukes of many high or low rolls.

Materials List/Setup

Station 1   two number cubes (1–6)

Station 2   none

Station 3   number cube (1–6)

Station 4   number cube (1–6)

Discussion Guide

To support students in reflecting on the activities and to gather some formative information about student learning, use the following prompts to facilitate a class discussion to "debrief" the station activities.

Prompts/Questions

1. What conclusions can you draw from a scatter plot if all the points are in a line? What if the points are randomly distributed?

2. Why is it important to collect a lot of data when conducting research?

3. What is an example of a real-life situation where we might want to conduct a survey to answer a question?

4. How does conducting an experiment help us answer questions?

Think, Pair, Share

Have students jot down their own responses to questions, then discuss with a partner (who was not in their station group), and then discuss as a whole class.

Suggested Appropriate Responses

1. If the points are in a line, there is a correlation between the two variables. If the points are randomly distributed, one variable does not affect the other.

2. because if you don't collect enough data, you won't notice the trends

3. anything where we care what a specific population does, wants, etc., such as how much money most people spend at lunch

4. It lets us see how frequently we get different outcomes and what those outcomes are.

Possible Misunderstandings/Mistakes

- Not asking questions that can be answered by collecting data
- Developing questions that are too broad (e.g., What is your name?)
- Not properly labeling the graph
- Confusing the axes

Statistics and Probability
Set 1: Collecting, Organizing, and Analyzing Data

Station 1

At this station, you will find two number cubes. You will be using these number cubes to generate data which you will then graph.

You will roll the number cubes one at a time. Record the number that comes up on the first number cube. Roll the second number cube, and record the sum of the two number cubes. Do this 30 times and record the data in the table below.

First number cube	Second number cube	Sum of number cubes	First number cube	Second number cube	Sum of number cubes

Using this data, create a scatter plot on a separate sheet of paper. Be sure to title the scatter plot and use appropriate labels.

What is the range of the sums? _____

What do you notice about your scatter plot? Is there a general trend?

_____

_____

_____

Statistics and Probability
Set 1: Collecting, Organizing, and Analyzing Data

Station 2

At this station, you will come up with different questions you could answer by collecting data.

Work with your group to come up with five questions you could answer by going out and collecting data. You need to be able to collect at least 30 pieces of data in order to answer any of these questions. (For example, what is the average number of siblings students in our class have?)

_____

_____

_____

_____

_____

Choose one of those questions to focus on.

How would you gather data to answer your question? Make sure to gather at least 30 pieces of data.

_____

_____

What materials would you need to gather the data? (e.g., number cubes)

_____

_____

How would you present your data? (e.g., bar graph, table)

_____

_____

Why is it important that you collect at least 30 pieces of data? Why are 5 pieces not enough?

_____

_____

Statistics and Probability
Set 1: Collecting, Organizing, and Analyzing Data

Station 3

At this station, you will try to determine which side of a number cube comes up most often.

Each member of your group should roll a number cube 25 times, keeping track of how many times each number comes up.

Number rolled	Tally	Final number
1		
2		
3		
4		
5		
6		

Combine your data in the table below.

Number rolled	Total
1	
2	
3	
4	
5	
6	

Based on this data, which number do you think has the best chance of coming up? _____

Some people say that they roll a 1 more often than a 6. Does your data support that statement? Explain.

_____

_____

_____

What are two questions you could answer with this data?

_____

_____

_____

Statistics and Probability
Set 1: Collecting, Organizing, and Analyzing Data

Station 4

At this station, you will find a number cube.

Roll the number cube 5 times and record your results in the table below.

Number rolled	Tally	Final number
1		
2		
3		
4		
5		
6		

What is the value of the mean roll? _____

Now roll the number cube 10 times. Record your results below.

Number rolled	Tally	Final number
1		
2		
3		
4		
5		
6		

What is the value of the mean roll? _____

continued

Statistics and Probability

Set 1: Collecting, Organizing, and Analyzing Data

Finally, roll the number cube 30 times. Record your results below.

Number rolled	Tally	Final number
1		
2		
3		
4		
5		
6		

As a group, review your data and answer the following questions.

What is the value of the mean roll? _____

How do the three means compare? What do you notice? _____

_____

_____

The mean value of the number cube is 3.5. Which mean is closest to that value? Why?

_____

_____

_____

Statistics and Probability

Goal: To provide opportunities for students to develop concepts and skills related to theoretical probability

Common Core State Standards

Statistics and Probability

Investigate chance processes and develop, use, and evaluate probability models.

7.SP.5. Understand that the probability of a chance event is a number between 0 and 1 that expresses the likelihood of the event occurring. Larger numbers indicate greater likelihood. A probability near 0 indicates an unlikely event, a probability around 1/2 indicates an event that is neither unlikely nor likely, and a probability near 1 indicates a likely event.

7.SP.6. Approximate the probability of a chance event by collecting data on the chance process that produces it and observing its long-run relative frequency, and predict the approximate relative frequency given the probability.

Student Activities Overview and Answer Key

Station 1

Students will look at two pairs of number cubes. They will compare the probability of rolling particular sums. In the end, they will draw conclusions about the two pairs of number cubes.

Answers: $1/6$; $5/36$; $1/6$; $5/36$; The probability of rolling each sum is the same for both pairs of number cubes.

Station 2

Students will consider two different spinners. They determine the theoretical probability of spinning various colors. They then work toward determining the probability of having one land on one color and the other land on another color, and discuss their strategy for this task.

Answers: $1/4$; $1/4$; $1/2$; $1/2$; $1/8$; using a tree diagram, writing out all the possibilities, multiplying 4×2

Station 3

Students will use a tree diagram to show the possible combinations when rolling two number cubes. They will use their diagram to answer questions about the likeliness of rolling different sums.

Answers: 11; no, because some sums repeat themselves; rolling a 7: the sum is 7 six times.

Station 4

Students find the theoretical probability for drawing cubes out of a bag. They then consider how the probabilities would be affected if everything in the bag was doubled, and draw conclusions.

Answers: $1/5$; $1/5$; $2/5$; because there are twice as many yellow cubes as any other color; $2/10$ or $1/5$; $4/10$ or $2/5$; the probabilities do not change

Materials List/Setup

Station 1   two number cubes with the following sides:
(4, 2, 2, 1, 3, 3) and (6, 1, 3, 5, 4, 8)

Station 2   two spinners; the first one should be divided evenly into red, green, blue, and yellow sections; the other should be evenly divided into black and white sections

Station 3   ruler
optional: two number cubes (1–6)

Station 4   bag containing one each of red, blue, and green cubes, and two yellow cubes

Discussion Guide

To support students in reflecting on the activities, and to gather formative information about student learning, use the following prompts to facilitate a class discussion to "debrief" the station activities.

Prompts/Questions

1. In what real-life situation might it be good to know the theoretical probability of an event happening?

2. You have 10 blocks in a bag—5 red, 2 yellow, 1 blue, and 2 green—and have figured out the probability of drawing a red block. Now you triple the number of blocks in the bag (15 red, 6 yellow, 3 blue, and 6 green). How does this affect the theoretical probability?

3. Make a general statement about what happens to the probability of an event happening (pulling out a red block, for example) if we multiply each choice by the same number.

4. How can tree diagrams help you determine theoretical probabilities?

Think, Pair, Share

Have students jot down their own responses to questions, discuss their responses with a partner (who was not in their station group), and then discuss as a whole class.

Suggested Appropriate Responses

1. when buying a lottery ticket—you might decide not to since the odds are against you

2. It does not affect it.

3. The probability does not change.

4. They show you all the possible outcomes.

Possible Misunderstandings/Mistakes

*   Not understanding that the ratio $2/5$ is equivalent to $4/10$

*   Miscounting the number of answers that satisfy the probability you are looking for

*   Flipping the ratios (e.g., $5/2$ instead of $2/5$)

Statistics and Probability
Set 2: Theoretical Probability

Station 1

At this station, you will find a very special pair of number cubes.

The table below shows the values on each number cube. Fill in the possible sums.

	1	2	2	3	3	4
1						
3						
4						
5						
6						
8						

What is the theoretical probability of rolling a 6? Write your solution as a ratio. _____

What is the theoretical probability of rolling an 8? Write your solution as a ratio. _____

Below is a table of all the sums possible when rolling two regular number cubes (sides 1–6).

	1	2	3	4	5	6
1	2	3	4	5	6	7
2	3	4	5	6	7	8
3	4	5	6	7	8	9
4	5	6	7	8	9	10
5	6	7	8	9	10	11
6	7	8	9	10	11	12

What is the theoretical probability of rolling a 6? Write your solution as a ratio. _____

What is the theoretical probability of rolling an 8? Write your solution as a ratio. _____

Look at both tables closely. What do you notice? _____

_____

_____

Statistics and Probability
Set 2: Theoretical Probability

Station 2

At this station, you will find two spinners. One spinner is evenly divided into four colors—red, green, blue, and yellow.

Work with your group to determine the following probabilities. Write them as a ratio.

P(yellow) = _____

P(red) = _____

The second spinner is evenly divided into two colors—black and white.

Work with your group to determine the following probabilities. Write them as a ratio.

P(black) = _____

P(white) = _____

What is the probability of spinning yellow and white? Use a tree diagram to draw out the different options, if necessary.

What strategy did you use to determine the probability of spinning yellow and white?

_____

_____

Statistics and Probability
Set 2: Theoretical Probability

Station 3

At this station, you will use a tree diagram to help you figure out the theoretical probability of different totals you could get when rolling two number cubes.

Draw a tree diagram to show all the possible combinations of rolls if your number cubes are both numbered 1–6.

Write the probability of getting each total as a ratio on your tree diagram.

How many different possible totals can you get by rolling two number cubes? _____

Are all the totals equally likely? How do you know? _____

_____

_____

Which outcome is most likely? How do you know? _____

_____

Statistics and Probability

Set 2: Theoretical Probability

Station 4

At this station, you will be working on determining theoretical probability. You will see a bag with five cubes in it—one green, one blue, one red, and two yellow.

You should imagine what would happen if you pulled out one cube. Find the following theoretical probabilities. Be sure to write the probabilities as ratios.

P(green) = _____

P(blue) = _____

P(yellow) = _____

Why is the probability of pulling out a yellow different from the probability of drawing another color?

_____

_____

What would the probability of drawing a red be if we doubled the blocks that were in the bag (2 green, 2 blue, 2 red, and 2 yellow)? _____

What would the probability be of drawing a yellow in this same situation? _____

What do you notice about the probabilities if we double everything in the bag?

_____

_____

_____

Statistics and Probability

Set 3: Experimental Probability

Instruction

Goal: To provide opportunities for students to develop concepts and skills related to understanding experimental probability and theoretical probability

Common Core State Standards

Statistics and Probability

Investigate chance processes and develop, use, and evaluate probability models.

> **7.SP.6.** Approximate the probability of a chance event by collecting data on the chance process that produces it and observing its long-run relative frequency, and predict the approximate relative frequency given the probability.

> **7.SP.7.** Develop a probability model and use it to find probabilities of events. Compare probabilities from a model to observed frequencies; if the agreement is not good, explain possible sources of the discrepancy.

> **a.** Develop a uniform probability model by assigning equal probability to all outcomes, and use the model to determine probabilities of events.

> **b.** Develop a probability model (which may not be uniform) by observing frequencies in data generated from a chance process.

Student Activities Overview and Answer Key

Station 1

Students will predict the number of times they will pull a red, green, and blue block from a bag. They will use theoretical probability to make this prediction. Students will then draw a cube from the bag and replace it 36 times in intervals of 12. They will compare their experimental results along the way with the theoretical results, and draw conclusions.

Answers: First table—4, 8, 12; second set of tables—answers vary depending on what blocks are pulled out; answers vary depending on what blocks are pulled out; answers may vary, but most likely students will be closer to their prediction for 36 turns than for 12 turns; Yes, the more turns we have, the closer the experimental probability is to the theoretical probability.

Station 2

Students will work with a list of randomly generated numbers. They predict what the mean should be based on the experimental probability of getting each number. They then compare their prediction with the actual average in a series of intervals. They will compare their experimental results with the theoretical result, and draw conclusions.

Answers: 5; table—answers will vary; The mean should approach the prediction as the amount of numbers increases; The experimental mean should be 5.

Station 3

Students will find the theoretical probability for rolling a 5 on a number cube. They will then each roll the number cube 20 times and compare the experimental probability to the theoretical probability. They will repeat this comparison after combining group data. Students then draw conclusions about sample size and experimental probability.

Answers: $\frac{1}{6}$; answers will vary; answers will vary; answers will vary—probably the second number because there was a larger sample size, and as sample size increases, experimental probability approached theoretical probability

Station 4

Students individually flip a coin 50 times to generate data. They then combine their data and look at the percentage of heads and tails that come up. Students discuss what the breakdown would look like if they flipped the coin 100,000 times.

Answers: Answers will vary; answers will vary; about 50% each because the experimental probability approaches the theoretical probability when sample sizes are very large

Materials List/Setup

Station 1   three blocks (red, green, and blue); a bag

Station 2   list of 100 randomly generated numbers (1–9)

Station 3   number cubes for each group member (1–6)

Station 4   coin for each group member

Discussion Guide

To support students in reflecting on the activities, and to gather formative information about student learning, use the following prompts to facilitate a class discussion to "debrief" the station activities.

Prompts/Questions

1. Why do we need a large sample size for the experimental probability to approach the theoretical probability?

2. Why do small samples differ from the theoretical probability?

3. Why is it important that we use random numbers or randomly pull a block out of the bag?

4. If you flipped a coin 1,000,000 times, how many times do you think heads would come up?

Think, Pair, Share

Have students jot down their own responses to questions, discuss their responses with a partner (who was not in their station group), and then discuss as a whole class.

Suggested Appropriate Responses

1. Large sample sizes account for any unusual outcomes that small sample sizes do not (e.g., rolling five 6s in a row).

2. Small samples can be affected by oddities in data.

3. Theoretical probability is not biased, so we should try to eliminate bias from experimental probability.

4. about 500,000 times

Possible Misunderstandings/Mistakes

- Difficulty dealing with a large amount of data

- Expecting small samples to look like the theoretical probability

- Thinking that when they have 30 or more pieces of data, the experimental probability will look exactly like the theoretical probability

Statistics and Probability
Set 3: Experimental Probability

Station 1

At this station, you will find three blocks in a bag—one red, one green, and one blue. Imagine that you pull out a block, record what color it is, and put it back in.

Work with your group to determine how many times you would expect to pull out the green block if you went through this routine 12 times, 24 times, and 36 times. This should be the theoretical probability. Record your predictions in the table below.

Number of times you choose a block	Predicted number of green blocks you will choose
12	
24	
36	

Now we are going to test our predictions. Place the three blocks in the bag. Without looking, take one block out. Record the color of the block you pulled out in the table below. Also, total the number of green blocks you pulled out after every 12 turns. Then answer the questions on the next page.

Turn	1	2	3	4	5	6	7	8	9	10	11	12	Total green 1–12	Total green 1–12
Color														

Turn	13	14	15	16	17	18	19	20	21	22	23	24	Total green 13–24	Total green 1–24
Color														

Turn	25	26	27	28	29	30	31	32	33	34	35	36	Total green 25–36	Total green 1–36
Color														

continued

Statistics and Probability
Set 3: Experimental Probability

How close were your original predictions? _____

_____

Were you closer with your guess about how many green blocks would be pulled after 12 turns or after 36 turns?

_____

Do you think you would be even closer if you had to estimate how many green blocks you would pull out after 300 turns? Why or why not?

_____

_____

_____

Statistics and Probability
Set 3: Experimental Probability

Station 2

At this station, you have a list of 100 randomly generated numbers. Each number, 1–9, has an equally likely chance of coming up.

If the experimental probability of a number coming up is $\dfrac{1}{9}$ for every number, what do you think the mean of your list of numbers will be?

_____

Use the list of randomly generated numbers to fill in the table below. Under "Sum," write the total sum of the first 25, 50, 75, and 100 numbers. For "Mean," determine the mean of the set of data up to that point.

Amount of numbers	Sum	Mean
25		
50		
75		
100		

What do you notice about your mean prediction as the amount of numbers you look at increases?

_____

_____

What do you think would happen if you found the mean of 10,000 randomly generated numbers (1–9)? Explain.

_____

_____

Statistics and Probability
Set 3: Experimental Probability

Station 3

At this station, you will find enough number cubes for each group member. Each group member should take one number cube. Work as a group to answer the following question.

What is the probability of rolling a 5? _____

Now each group member will roll a number cube 20 times on his or her own, and record the rolls in the table below.

Number rolled	Tally	Total number
1		
2		
3		
4		
5		
6		

What was the experimental probability of rolling a 5? _____

Combine data with your group.

Number rolled	Total number
1	
2	
3	
4	
5	
6	

What is the experimental probability of rolling a 5? _____

Which experimental probability was closer to the theoretical probability? Why?

_____

_____

Statistics and Probability
Set 3: Experimental Probability

Station 4

You will find a coin for every group member at this station.

Each group member should flip the coin 50 times, and record his or her data in the table below.

Result	Tally	Total number
Heads		
Tails		

Now combine data with your group members.

Result	Total number
Heads	
Tails	

What percentage of the flips was heads? _____

What percentage of the flips was tails? _____

What do you think the percentages would be if you flipped the coin 100,000 times? Why?

_____

_____